RENEWABLE ENERGY: RESEARCH, DEVELOPMENT AND POLICIES SERIES

BIOLOGICAL BARRIERS TO CELLULOSIC ETHANOL

RENEWABLE ENERGY: RESEARCH, DEVELOPMENT AND POLICIES SERIES

Renewable Fuel Standard Issues
Daniel T. Crowe (Editor)
2009. ISBN: 978-1-60692-289-7

Ethanol and Biofuels: Production, Standards and Potential
Wesley P. Leland (Editor)
2009 ISBN: 978-1-60692-224-8

Ethanol and Biofuels: Production, Standards and Potential
Wesley P. Leland (Editor)
2009 ISBN: 978-1-60876-592-8 (Online Book)

Wind Power: Technology, Economics and Policies
Cedrick N. Osphey (Editor)
2009. ISBN :978-1-60692-323-8

Solar Collectors: Energy Conservation, Design and Applications
Arthur V. Killian (Editor)
2009. ISBN: 978-1-60741-069-0

Solar Collectors: Energy Conservation, Design and Applications
Arthur V. Killian (Editor)
2009. ISBN: 978-1-60876-920-9 (Online book)

Wind Energy in Electricity Markets with High Wind Penetration
Julio Usaola and Edgardo D. Castronuovo
2009. ISBN: 978-1-60741-153-6

Renewable Energy Grid Integration: The Business of Photovoltaics
Marco H. Balderas (Editor)
2009. ISBN: 978-1-60741-324-0

Biomass Gasification: Chemistry, Processes and Applications
Jean-Pierre Badeau and Albrecht Levi (Editors)
2009. ISBN: 978-1-60741-461-2

Renewable Energies: Feasibility, Time and Cost Options
John O'M. Bockris
2009. ISBN: 978-1-60876-006-0

Biological Barriers to Cellulosic Ethanol
Ernest V. Burkheisser
2010. ISBN: 978-1-60692-203-3

Solar America: How, What and When?
Nash M. Perales (Editor)
2010. ISBN: 978-1-60741-333-2

RENEWABLE ENERGY: RESEARCH, DEVELOPMENT
AND POLICIES SERIES

BIOLOGICAL BARRIERS TO CELLULOSIC ETHANOL

ERNEST V. BURKHEISSER
EDITOR

Nova Science Publishers, Inc.
New York

Copyright © 2010 by Nova Science Publishers, Inc.

All rights reserved. No part of this book may be reproduced, stored in a retrieval system or transmitted in any form or by any means: electronic, electrostatic, magnetic, tape, mechanical photocopying, recording or otherwise without the written permission of the Publisher.

For permission to use material from this book please contact us:
Telephone 631-231-7269; Fax 631-231-8175
Web Site: http://www.novapublishers.com

NOTICE TO THE READER

The Publisher has taken reasonable care in the preparation of this book, but makes no expressed or implied warranty of any kind and assumes no responsibility for any errors or omissions. No liability is assumed for incidental or consequential damages in connection with or arising out of information contained in this book. The Publisher shall not be liable for any special, consequential, or exemplary damages resulting, in whole or in part, from the readers' use of, or reliance upon, this material. Any parts of this book based on government reports are so indicated and copyright is claimed for those parts to the extent applicable to compilations of such works.

Independent verification should be sought for any data, advice or recommendations contained in this book. In addition, no responsibility is assumed by the publisher for any injury and/or damage to persons or property arising from any methods, products, instructions, ideas or otherwise contained in this publication.

This publication is designed to provide accurate and authoritative information with regard to the subject matter covered herein. It is sold with the clear understanding that the Publisher is not engaged in rendering legal or any other professional services. If legal or any other expert assistance is required, the services of a competent person should be sought. FROM A DECLARATION OF PARTICIPANTS JOINTLY ADOPTED BY A COMMITTEE OF THE AMERICAN BAR ASSOCIATION AND A COMMITTEE OF PUBLISHERS.

LIBRARY OF CONGRESS CATALOGING-IN-PUBLICATION DATA

Burkheisser, Ernest V.
Biological barriers to cellulosic ethanol / Ernest V. Burkheisser.
   p. cm.
Includes index.
ISBN 978-1-60692-203-3 (softcover)
1. Cellulose--Biotechnology. 2. Biomass energy. I. Title.
TP248.65.C45B87 2009

Published by Nova Science Publishers, Inc. ✦ *New York*

TABLE OF CONTENTS

Preface		ix
Chapter 1	Introduction *United States Department of Energy*	1
Chapter 2	Technical Strategy: Development of a Viable Cellulosic Biomass to Biofuel Industry *United States Department of Energy*	31
Chapter 3	Systems Biology to Overcome Barriers to Cellulosic Ethanol *United States Department of Energy*	39
Chapter 4	Feedstocks for Biofuels *United States Department of Energy*	59
Chapter 5	Deconstructing Feedstocks to Sugars *United States Department of Energy*	89
Chapter 6	Sugar Fermentation to Ethanol *United States Department of Energy*	125
Chapter 7	Crosscutting 21st Century Science, Technology, and Infrastructure for a New Generation of Biofuel Research *United States Department of Energy*	161
Chapter 8	Bioprocess Systems Engineering and Economic Analysis *United States Department of Energy*	187
Chapter 9	Appendix A. Provisions for Biofuels and Biobased Products in the Energy Policy Act of 2005 *United States Department of Energy*	191
Chapter 10	Appendix B. Workshop Participants *United States Department of Energy*	195
Chapter 11	Appendix C. Workshop Participant Biosketches *United States Department of Energy*	205
Chapter Sources		233
Index		235

PREFACE

The purpose of this book is to define barriers and challenges to a rapid expansion of cellulosic-ethanol production and determine ways to speed solutions through concerted application of modern biology tools as part of a joint research agenda. Although the focus was ethanol, the science applies to additional fuels that include biodiesel and other bioproducts or coproducts having critical roles in any deployment scheme. The core barrier is cellulosic-biomass recalcitrance to processing to ethanol. Biomass is composed of nature's most ready energy source, sugars, but they are locked in a complex polymer composite exquisitely created to resist biological and chemical degradation. Key to energizing a new biofuel industry based on conversion of cellulose (and hemicelluloses) to ethanol is to understand plant cell-wall chemical and physical structures—how they are synthesized and can be deconstructed. With this knowledge, innovative energy crops—plants specifically designed for industrial processing to biofuel—can be developed concurrently with new biology-based treatment and conversion methods. Recent advances in science and technological capabilities, especially those from the nascent discipline of systems biology, promise to accelerate and enhance this development. Resulting technologies will create a fundamentally new process and biorefinery paradigm that will enable an efficient and economic industry for converting plant biomass to liquid fuels. These key barriers and suggested research strategies to address them are described in this book.

The core barrier is cellulosic-biomass recalcitrance to processing to ethanol. Biomass is composed of nature's most ready energy source, sugars, but they are locked in a complex polymer composite exquisitely created to resist biological and chemical degradation. Key to energizing a new biofuel industry based on conversion of cellulose (and hemicelluloses) to ethanol is to understand plant cell-wall chemical and physical structures—how they are synthesized and can be deconstructed. With this knowledge, innovative energy crops—plants specifically designed for industrial processing to biofuel—can be developed concurrently with new biology-based treatment and conversion methods. Recent advances in science and technological capabilities, especially those from the nascent discipline of systems biology, promise to accelerate and enhance this development. Resulting technologies will create a fundamentally new process and biorefinery paradigm that will enable an efficient and economic industry for converting plant biomass to liquid fuels. These key barriers and suggested research strategies to address them are described in this book.

Chapter 1 - In the 2006 Union address the president Bush outlined the new Advanced Energy Initiative (AEI) to help overcome America's dependence on foreign sources of energy (AEI 2006) and the American Competitiveness Initiative to increase R&D investments and strengthen education (ACI 2006). He seeks to reduce our national dependence on imported oil by accelerating the development of domestic, renewable alternatives to gasoline and diesel fuels.

Breakthrough technologies to realize the potential of cellulosic biofuels can be expedited by application of a new generation of biological research created by the genome revolution. Overcoming barriers to development of these fuels on an industrial scale will require high-performance energy feedstocks and microbial processes, both to break down feedstocks to sugars and to ferment sugars to ethanol. A focused set of investments linking revolutionary biofuel technologies with advances from the biological, physical, computational, and engineering sciences will quickly remove barriers to an efficient, economic, and sustainable biofuel industry.

Chapter 2 - Innovative energy crops—plants specifically designed for industrial processing to biofuels—can be developed concurrently with new biorefinery treatment and conversion processes. Recent advances in science and capabilities, especially those from the nascent discipline of systems biology, promise to accelerate and enhance this paradigm. Resulting technologies will allow the fusion of agriculture, industrial biotechnology, and energy value chains to enable an efficient and economically viable industry for conversion of plant biomass to liquid fuels. Displacing up to 30% of the nation's current transportation fuel by 2030 will require significant policy support and technical advancement over the next 5 to 15 years. Research and technology development described in this roadmap will occur in three phases to enable industry to meet the 2025 and 2030 goals.

Chapter 3 - Lignocellulosic biomass has long been recognized as a potential low-cost source of mixed sugars for fermentation to fuel ethanol. Plant biomass has evolved effective mechanisms for resisting assault on its structural sugars from the microbial and animal kingdoms. This property underlies a natural recalcitrance, creating technical barriers to the cost-effective transformation of lignocellulosic biomass to fermentable sugars. Moderate yields and the resulting complex composition of sugars and inhibitory compounds lead to high processing costs. Several technologies have been developed over the past 80 years, often in wartime, that allow this conversion process to occur, yet the clear objective now is to make the process cost-competitive in today's markets.

Chapter 4 - One critical foundation for developing bioenergy crops and their processing technologies is ameliorating cell-wall recalcitrance to breakdown. Understanding cell walls is essential for optimizing their synthesis and the processes used to deconstruct them to sugars for conversion to ethanol. A prerequisite for a competitive biofuel industry is the development of crops that have both desirable cell-wall traits and high biomass productivity under sustainable low-input conditions. Major agricultural crops grown today for food, feed, and fiber in the United States have not been bred for biofuels. Thus, many carefully selected traits in food and feed crops, such as a high ratio of seed to straw production (harvest index), are disadvantageous in biofuel production. A suite of new crops and new varieties of existing crops specifically bred for biofuels and adapted to a range of different soil types and climatic conditions is required.

Chapter 5 - This chapter describing the challenges of deconstructing cellulosic biomass to ethanol is critically linked to both the feedstock and fermentation areas. Lignocellulose recalcitrance to bioprocessing will remain the core problem and will be the limiting factor in creating an economy based on lignocellulosic ethanol production. Understanding biomass recalcitrance will help to drive crop design. Knowledge about feedstock breakdown mechanisms and products will drive fermentation and consolidation strategies, which ultimately will include consolidated bioprocessing (CBP). CBP incorporates the traits for deconstruction and fermentation of sugars to ethanol into a single microbe or culture. The previous chapter describes how tools of modern genomics-based systems biology can provide tremendous opportunities to engineer energy-plant genomes for new varieties. Those engineered plants will grow more efficiently while also producing optimal polysaccharide compositions for deconstruction to sugars and fermentation to ethanol and other products. Further advancements in plant engineering also can generate new energy crops and trees harboring polysaccharide storage structures (principally in the cell walls) that are *designed* for deconstruction. This achievement will be an important outcome of scientific research needed to optimize deconstruction of native cell walls in such crop residues as corn stover and wheat straw and such energy crops as switchgrass and poplar.

Chapter 6 - Fermentation of sugars by microbes is the most common method for converting sugars inherent within biomass feedstocks into liquid fuels such as ethanol. Bioconversion or biocatalysis is the use of microbes or enzymes to transform one material into another. The process is well established for some sugars, such as glucose from cornstarch, now a mature industry. Production of fuel ethanol from the mixture of sugars present in lignocellulosic biomass, however, remains challenging with many opportunities for improvement. More robust microorganisms are needed with higher rates of conversion and yield to allow process simplification through consolidating process steps. This development would reduce both capital and operating costs, which remain high by comparison with those of corn.

Chapter 7 - Efficiently and inexpensively producing ethanol or alternative products such as alkanes, fatty acids, and longer-chain alcohols from biomass will require significant advances in our understanding and capabilities in three major areas explored at this workshop: Feedstocks for Biofuels, Deconstructing Feedstocks to Sugars, and Sugar Fermentation to Ethanol. A systems-level approach to understanding and manipulating plants and microorganisms central to processing biomass into liquid fuels depends on obtaining and using detailed chemical and biochemical information on organism states and structures to build functional models that guide rational design and engineering. A systems-level understanding of model plants will facilitate rational improvement of plant cell-wall composition in crops dedicated to conversion into biofuels. New approaches and tools will be necessary to characterize definitively the detailed organizational structures of principal types of plant cellulose and their relative energies and interrelationships with such other structural components as lignins and noncellulosic polysaccharides.

Chapter 8 - Complete process modeling should be initiated to guide the scientific work described in this Roadmap, including systems engineering and economic analyses, to evaluate the most-probable scenarios; to coordinate advances and needed research across the feedstock, deconstruction, and fermentation domains; and to reduce risk as the development cycle proceeds. Results and methodologies should be made available to the community through the web and should be subject to continuous improvements based on community feedback.

Biomass-conversion literature has many examples of pretreatment, hydrolysis, and fermentation systems that are technically effective but have no real chance to be competitive economically. A common tendency among biotechnologists is to attempt processes that will achieve, for example, the highest yields, rates, and titers. Process modeling, however, very often reveals that different combinations of these parameters are needed to produce a commercially viable process. Disciplined systems engineering and economic analysis using mass balances and standard analytical methods can eliminate ineffective approaches rather quickly, narrowing the focus to the most-promising options. Current advanced saccharification enzyme systems have demonstrated what a concerted, focused program of enzyme development can achieve. These systems, however, were designed for a specific acid-based pretreatment and the biomass raw material of corn stover. Other combinations of pretreatments and biomass materials will need to be analyzed and subjected to process engineering and economic analyses as they mature.

In: Biological Barriers to Cellulosic Ethanol
Editor: Ernest V. Burkheisser

ISBN: 978-1-60692-203-3
© 2010 Nova Science Publishers, Inc.

Chapter 1

INTRODUCTION

United States Department of Energy

In his 2006 State of the Union address (Bush 2006), the president outlined the new Advanced Energy Initiative (AEI) to help overcome America's dependence on foreign sources of energy (AEI 2006) and the American Competitiveness Initiative to increase R&D investments and strengthen education (ACI 2006). He seeks to reduce our national dependence on imported oil by accelerating the development of domestic, renewable alternatives to gasoline and diesel fuels.

> *"With America on the verge of breakthroughs in advanced energy technologies, the best way to break the addiction to foreign oil is through new technologies."* —White House Press Release on the State of the Union Address and AEI (January 31, 2006)

Breakthrough technologies to realize the potential of cellulosic biofuels can be expedited by application of a new generation of biological research created by the genome revolution. Overcoming barriers to development of these fuels on an industrial scale will require high-performance energy feedstocks and microbial processes, both to break down feedstocks to sugars and to ferment sugars to ethanol. A focused set of investments linking revolutionary biofuel technologies with advances from the biological, physical, computational, and engineering sciences will quickly remove barriers to an efficient, economic, and sustainable biofuel industry.

JOINT WORKSHOP CHALLENGES BIOFUEL SCIENCE AND TECHNOLOGY COMMUNITIES

Two Department of Energy (DOE) offices are teaming to advance biofuel development and use: The Office of Biological and Environmental Research (OBER) within the Office of Science (SC) and the Office of the Biomass Program (OBP) within the Office of Energy Efficiency and Renewable Energy (EERE) (see descriptions of the two DOE programs, pp. 17 and 19). These offices are challenging their communities to identify critical science needs to support a substantial and sustainable expansion of biomass-derived fuels, specifically cellulosic ethanol. In the jointly sponsored Biomass to Biofuels Workshop held December 7–9, 2005, in Rockville, Maryland, more than 50 scientists representing a wide range of expertise convened to define barriers and challenges to this new biofuel industry. The workshop concentrated on improvement of biomass crops and their processing to transportation fuels. Although the focus was ethanol, the science applies to additional fuels that include biodiesel and to other bioproducts or coproducts having critical roles in any deployment scheme.

The current approach to introducing biofuels relies on an "evolutionary" business and economic driver for a steady but moderate entry into the market. Technologies for implementing this new industry are being tested either by producing higher-value products from renewables (such as lactic acid) or as incremental additions to current corn-ethanol refineries (such as the conversion of residual corn-kernel fibers to ethanol).

This report is a workshop-produced roadmap for accelerating cellulosic ethanol research, helping make biofuels practical and cost-competitive by 2012 ($1.07/gal ethanol) and offering the potential to displace up to 30% of the nation's current gasoline use by 2030. It argues that rapidly incorporating new systems biology approaches via significant R&D investment will spur use of these technologies for expanded processing of energy crops and residues. Furthermore, this strategy will decrease industrial risk from use of a first-of-a-kind technology, allowing faster deployment with improved methods. Ultimately, these approaches foster setting more aggressive goals for biofuels and enhance the strategy's sustainability.

AMERICA'S ENERGY CHALLENGES

The triple energy-related challenges of the 21st Century are economic and energy growth, energy security, and climate protection. The United States imports about 60% of the petroleum it consumes, and that dependency is increasing.* Since the U.S. economy is tied so closely to petroleum products and oil imports, disruptions in oil supplies can result in severe economic and social impacts. Conventional oil production will peak in the near future, and the resulting energy transition will require a portfolio of responses, including unconventional fossil resources and biofuels. Environmental quality and climate change due to energy emissions are additional concerns. Annual U.S. transportation emissions of the greenhouse gas (GHG) carbon dioxide (CO_2) are projected to increase from about 1.9 billion metric tons in 2004 to about 2.7 billion metric tons in 2030 (EIA 2006).

The Promise of Biofuels

Fuels derived from cellulosic biomass**—the fibrous, woody, and generally inedible portions of plant matter—offer an alternative to conventional energy sources that supports national economic growth, national energy security, and environmental goals. Cellulosic biomass is an attractive energy feedstock because supplies are abundant domestically and globally. It is a renewable source of liquid transportation fuels that can be used readily by current-generation vehicles and distributed through the existing transportation-fuel infrastructure. Ethanol from corn grain is an increasingly important additive fuel source, but it has limited growth potential as a primary transportation fuel.*** The U.S. "starch-based" ethanol industry will jump start a greatly expanded ethanol industry that includes cellulosic ethanol as a major transportation fuel.

Cellulose and hemicelluloses, found in plant cell walls, are the primary component of biomass and the most plentiful form of biological material on earth. They are polysaccharides made up of energy-rich sugars that can be converted to ethanol (see sidebar, Understanding Biomass, p. 53). Current methods to break down biomass into simple sugars and convert them into ethanol are inefficient and constitute the core barrier to producing ethanol at quantities and costs competitive with gasoline.

Biological research is undergoing a major transformation. The systems biology paradigm—born of the genome revolution and based on high-throughput advanced technologies, computational modeling, and scientific-team approaches—can facilitate rapid progress and is a readily applicable model for biofuel technology. Systems biology is the core of the OBER Genomics:GTL program, whose goal is to achieve a predictive understanding of the complex network of interactions that underpin the biological processes related to biofuel production. Biological challenges to which GTL can apply systems biology approaches include enhancing the productivity of biomass crops optimized for industrial processing, improving enzyme systems that deconstruct plant cell walls, and increasing the yield of ethanol-producing microorganisms. Systems biology tools and knowledge will enable rational engineering of a new generation of bioenergy systems made up of sustainable energy crops for widely varying agroecosystems and tailored industrial processes. This research approach will encourage the critical fusion of the agriculture, industrial biotechnology, and energy sectors.

A GROWING MANDATE FOR BIOFUELS: POLICY, LEGISLATIVE, AND OTHER DRIVERS

A primary goal of the president's 2001 National Energy Policy (NEP) is to increase U.S. energy supplies, incorporating a more diverse mix of domestic resources to support growth in demand and to reduce national dependence on imported oil (NEPDG 2001). AEI accelerates and expands on several policy and legislative mandates (AEI 2006). It aims to reduce the nation's reliance on foreign oil in the near term and provides a 22% increase in clean-energy research at DOE for FY 2007, accelerating progress in renewable energy.

According to AEI, the United States must move beyond a petroleum-based economy and devise new ways to power automobiles. The country needs to facilitate domestic, renewable alternatives to gasoline and diesel fuels. The administration will accelerate research in cutting-edge methods of producing such "homegrown" renewable biobased transportation fuels as ethanol from agricultural and forestry feedstocks including wood chips, stalks, and switchgrass. AEI would foster the early commercialization of advanced biofuel technologies, enabling U.S. industry to lead in deploying biofuels and chemicals internationally.

Achieving the goal of displacing 30% of the nation's current gasoline use by 2030 would require production levels equal to roughly 60 billion gallons a year (Bgal/year) of ethanol (see Table 1. Comparisons of 2004 Gasoline and Ethanol Equivalents, this page). An annual supply of roughly a billion dry tons of biomass will be needed to support this level of ethanol production.A recent report by the U.S. Department of Agriculture (USDA) and DOE finds potential to sustainably harvest more than 1.3 billion metric tons of biomass from U.S. forest and agricultural lands by mid-21st Century (Perlack et al. 2005). Investments in R&D and infrastructure are needed to realize this feedstock potential.

The U.S. Energy Policy Act of 2005 (EPAct; Appendix A, Provisions for Biofuels and Biobased Products in the Energy Policy Act of 2005, p. 186) has established aggressive near-term targets for ethanol production. A key provision requires mixing 4 Bgal of renewable fuel with gasoline in 2006.This requirement increases annually to 7.5 Bgal of renewable fuel by 2012.For 2013 and beyond, the required volume will include a minimum of 250 million gallons (Mgal) of *cellulosic* ethanol. Another section of the EPAct authorizes funds for an incentive program to ensure the annual production of 1 Bgal of cellulosic biomass-based fuels by 2015. Ethanol is the most common biofuel produced from cellulose, but other possible biofuel compounds can be produced as well.

Other important legislative drivers supporting biofuels are the Biomass R&D Act of 2000 and Title IX of the Farm Bill 2002 (U.S. Congress 2000; U.S.Congress 2002).The Biomass R&D Act directed the departments of Energy and Agriculture to integrate their biomass R&D and established the Biomass Research and Development Technical Advisory Committee (BTAC), which advises the Secretary of Energy and the Secretary of Agriculture on strategic planning for biomass R&D. As a precedent to the current presidential initiative, in 2002 BTAC set a goal requiring biofuels to meet 20% of U.S. transportation fuel consumption by 2030 as part of its vision for biomass technologies (BTAC 2002). Title IX supports increased use of biobased fuels and products and incentives and grants for biofuel and biorefinery R&D.

In addition to legislative mandates, several independent studies have acknowledged the great potential of biofuels in achieving a more diverse domestic energy supply (NCEP 2004; Greene et al. 2004; Lovins et al. 2005). Growing support for developing biomass as a key energy feedstock is coming from a variety of national and international organizations (GEC 2005; Ag Energy Working Group 2004; IEA 2004). Although these reports differ in the amounts of gasoline that could be replaced by ethanol from biomass, they all agree on three key issues: (1) Current trends in energy use are not sustainable and are a security risk; (2) No single solution will secure the energy future—a diverse portfolio of energy options will be required; and (3) Biofuels can be a significant part of the transportation sector's energy solution.

In its evaluation of options for domestic production of motor fuels, the National Commission on Energy Policy (NCEP) recommended cellulosic biomass as an important topic for near-term federal research, development,and demonstration and found that "cellulosic ethanol has the potential to make a meaningful contribution to the nation's transportation fuel supply in the next two to three decades" (NCEP 2004).

The Natural Resources Defense Council (NRDC) has projected that an aggressive plan to develop cellulosic biofuels in the United States could "produce the equivalent of nearly 7.9 million barrels of oil per day by 2050 … more than 50 percent of our current total oil use in the transportation sector and more than three times as much as we import from the Persian Gulf alone" (Greene et al. 2004). This corresponds to roughly 100 Bgal/year ethanol. NRDC also recommends $1.1 billion in funding between 2006 and 2012 for biomass research, development, and demonstration with 45% of this funding focused on overcoming biomass recalcitrance to ethanol processing. This level of funding is expected to stimulate a regular flow of advances needed to make ethanol cost-competitive with gasoline and diesel.

An independent analysis from the Rocky Mountain Institute found that significant gains in energy efficiency and the large-scale displacement of oil with biofuels, mainly cellulosic ethanol, would be key components of its strategy to reduce American oil dependence over the next few decades (Lovins et al. 2005).

To illustrate the widespread support for fuel ethanol, the Governors' Ethanol Coalition, an organization devoted to the promotion and increased use of ethanol, now includes 32 member states as well as international representatives from Brazil, Canada, Mexico, Sweden, and Thailand. In a recent report, the coalition called for rapid expansion of ethanol to meet at least 10% of transportation fuel needs "as soon as practicable" and for development of "lignocellulosic-based" fuels for expansion beyond those levels (GEC 2005). "The use of ethanol, particularly biomass-derived ethanol, can produce significant savings in carbon dioxide emissions. This approach offers a no-regrets policy that reduces the potential future risks associated with climate change and has the added benefit of economic development."

BENEFITS OF BIOFUELS

Biofuels, especially corn-derived and cellulosic ethanol, constitute the only renewable liquid transportation fuel option that can be integrated readily with petroleum-based fuels, fleets, and infrastructure. Production and use of biofuels can provide substantial benefits to national energy security, economic growth, and environmental quality.

National Energy Security Benefits

"National security is linked to energy through the dependence of this country and many others on imported oil—much of it located in politically troubled parts of the globe. As such, the potential for large-scale failures in the global production and distribution system presents a real threat."
— Governors' Ethanol Coalition (GEC 2005)

Table 1. Comparisons of 2004 Gasoline and Ethanol Equivalents

2004	Gasoline (billion gallons)	Ethanol Equivalents (billion gallons)
U.S. consumption, 2004	139	200
About 60% from imports	83	120
Requirements to displace 30% of 2004 U.S. consumption	42	60
• Biomass requirements at 80 gal/ton		• 750 Mton
• Land requirements at 10 ton/acre and 80 gal/ton		• 75 Macre
• Numbers of refineries at 100 Mgal/refinery		• 600 (each requiring 160 miles^2 net or 125,000 acres)

Source: Adapted from ORNL *Review* (www.ornl.gov/info/ornlreview/v33i2i00/bioenergy.htm)

Figure 1. Reduced Carbon Dioxide Emissions of Ethanol from Biomass.
When compared with gasoline, ethanol from cellulosic biomass could dramatically reduce emissions of the greenhouse gas, carbon dioxide (CO_2). Although burning gasoline and other fossil fuels increases atmospheric CO_2 concentrations, the photosynthetic production of new biomass takes up most of the carbon dioxide released when bioethanol is burned.

Today the United States is dependent on oil for transportation. Alternative, domestically based, and sustainable fuel-development strategies, therefore, are essential to ensuring national security. America accounts for 25% of global oil consumption yet holds only 3% of the world's known oil reserves. About 60% of known oil reserves are found in sensitive and volatile regions of the globe. Increasing strain on world oil supply is expected as developing countries become more industrialized and use more energy. Any strategy to reduce U.S. reliance on imported oil will involve a mix of energy technologies including conservation. Biofuels are an attractive option to be part of that mix because biomass is a domestic, secure, and abundant feedstock. Global availability of biomass feedstocks also would provide an international alternative to dependence on an increasingly strained oil-distribution system as well as a ready market for biofuel-production technologies.

Economic Benefits

A biofuel industry would create jobs and ensure growing energy supplies to support national and global prosperity. In 2004, the ethanol industry created 147,000 jobs in all sectors of the economy and provided more than $2 billion of additional tax revenue to federal, state, and local governments (RFA 2005). Conservative projections of future growth estimate the addition of 10,000 to 20,000 jobs for every billion gallons of ethanol production (Petrulis 1993).

In 2005 the United States spent more than $250 billion on oil imports,and the total trade deficit has grown to more than $725 billion (U.S. Commerce Dept. 2006). Oil imports, which make up 35% of the total, could rise to 70% over the next 20 years (Ethanol Across America 2005).

Among national economic benefits, a biofuel industry could revitalize struggling rural economies. Bioenergy crops and agricultural residues can provide farmers with an important new source of revenue and reduce reliance on government funds for agricultural support. An economic analysis jointly sponsored by USDA and DOE found that the conversion of some cropland to bioenergy crops could raise depressed traditional crop prices by up to 14%. Higher prices for traditional crops and new revenue from bioenergy crops could increase net farm income by $6 billion annually (De La Torre Ugarte 2003).

Environmental Benefits

Climate Change

When fossil fuels are consumed, carbon sequestered from the global carbon cycle for millions of years is released into the atmosphere, where it accumulates. Biofuel consumption can release considerably less CO_2, depending on how it is produced. The photosynthetic production of new generations of biomass takes up the CO_2 released from biofuel production and use (see Figure. 1. Reduced Carbon Dioxide Emissions of Ethanol from Biomass, this page). A life-cycle analysis shows fossil CO_2 emissions from cellulosic ethanol to be 85% lower than those from gasoline (Wang 2005).These emissions arise from the use of fossil energy in producing cellulosic ethanol. Nonbiological sequestration of CO_2 produced by the fermentation process can make the biofuel enterprise net carbon negative.

A recent report (Farrell et al. 2006) finds that ethanol from cellulosic biomass reduces substantially both GHG emissions and nonrenewable energy inputs when compared with gasoline. The low quantity of fossil fuel required to produce cellulosic ethanol (and thus reduce fossil GHG emissions) is due largely to three key factors. First is the yield of cellulosic biomass per acre. Current corn-grain yields are about 4.5 tons/acre.Starch is 66% by weight, yielding 3 tons to produce 416 gal of ethanol,compared to an experimental yield of 10 dry tons of biomass/acre for switchgrass hybrids in research environments (10 dry tons at a future yield of 80 gal/ton = 800 gal ethanol). Use of corn grain, the remaining solids (distillers' dried grains), and stover could yield ethanol at roughly 700 gal/acre. Current yield for nonenergy-crop biomass resources is about 5 dry tons/acre and roughly 65 gal/ton. The goal for energy crops is 10 tons/acre at 80 to 100 gal/ton during implementation. Second, perennial biomass crops will take far less energy to plant and cultivate and will require less nutrient, herbicide, and fertilizer. Third, biomass contains lignin and other recalcitrant residues that can be burned to produce heat or electricity consumed by the ethanol-production process.

Energy crops require energy inputs for production, transportation, and processing—a viable bioenergy industry will require a substantial positive energy balance. Figure 2. Comparison of Energy Yields with Energy Expenditures, this page, compares results for cellulosic and corn ethanol, gasoline,and electricity, demonstrating a substantially higher yield for cellulosic ethanol. Over time a mature bioenergy economy will substitute biomass-derived energy sources for fossil fuel, further reducing net emissions.

Other Environmental Benefits

Perennial grasses and other bioenergy crops have many significant environmental benefits over traditional row crops (see Figure. 3. *Miscanthus* Growth over a Single Growing Season in Illinois, p. 9). Perennial energy crops provide a better environment for more-diverse wildlife habitation. Their extensive root systems increase nutrient capture, improve soil quality, sequester carbon, and reduce erosion. Ethanol, when used as a transportation fuel, emits less sulfur, carbon monoxide, particulates, and GHGs (Greene et al. 2004).

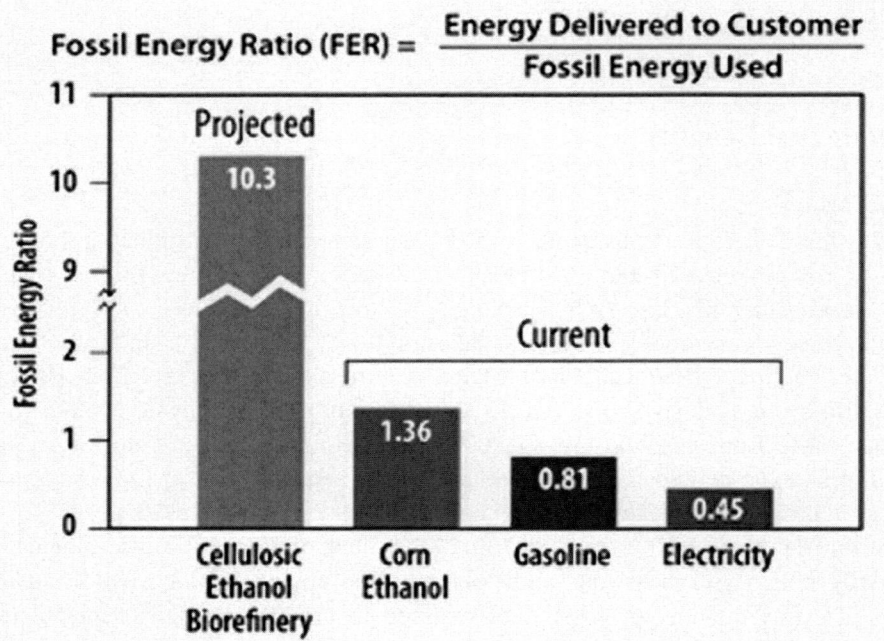

Source: Figure, based on the Argonne National Laboratory GREET model, is derived from Brinkman et al. 2005. Other papers that support this study include Farrell et al. 2006 and Hammerschlag 2006.

Figure 2. Comparison of Energy Yields with Energy Expenditures.
The fossil energy–replacement ratio (FER) compares energy yield from four energy sources with the amount of fossil fuel used to produce each source. Note that the cellulosic ethanol biorefinery's projected yield assumes future technological improvements in conversion efficiencies and advances that make extensive use of a biomass crop's noncellulosic portions for cogeneration of electricity. Similar assumptions would raise corn ethanol's FER if, for example, corn stover were to replace current natural gas usage. The corn ethanol industry, already producing ethanol as an important additive and fuel extender, is providing a foundation for expansion to cellulosic ethanol.

Image source: S. Long, University of Illinois

Figure 3. *Miscanthus* Growth over a Single Growing Season in Illinois. *Miscanthus* has been explored extensively as a potential energy crop in Europe and now is being tested in the United States. The scale is in feet. These experiments demonstrate results that are feasible in development of energy crops.

FEASIBILITY OF BIOFUELS

The United States could benefit substantially by increasing its use of domestic, renewable fuels in the transportation sector, but can biofuels be produced at the scale needed to make a real difference in transportation consumption of fossil fuels? More specifically, is there enough land to provide the needed large-scale supply of biomass, is the use of biofuels sustainable agriculturally, can biofuels become cost-competitive with gasoline, and is cellulosic-biofuel production technically feasible for energy? The short answer to all these questions is yes, and this section summarizes recent reports that support this view.

Land Availability

A major factor influencing the extent to which biofuels will contribute to America's energy future is the amount of land available for biomass harvesting. Are biomass resources sufficient to meet a significant portion of transportation-fuel consumption, and how would harvesting biomass for energy affect current agricultural and forestry practices?

In 2005, a study jointly supported by DOE and USDA examined whether land resources in the United States are sufficient to sustain production of over 1 billion dry tons of biomass annually, enough to displace 30% or more of the nation's current consumption of liquid transportation fuels.By assuming relatively modest changes in agricultural and forestry practices, this study projects that 1.366 billion dry tons of biomass could be available for large-scale bioenergy and biorefinery industries by mid-21st Century while still meeting demand for forestry products,food, and fiber (Perlack et al.2005) (see sidebar, A Billion-Ton Annual Supply of Biomass, p. 10). This supply of biomass would be a sevenfold increase over the 190 million dry tons of biomass per year currently used for bioenergy and bioproducts.Most of this biomass is burned for energy, with only 18 million dry tons used for biofuels (primarily corn-grain ethanol) and 6 million dry tons used for bioproducts.

The biomass potential determined by the "billion-ton" study is one scenario based on a set of conservative assumptions derived from current practices and should not be considered an upper limit. Crop-yield increases assumed in this study follow business-as-usual expectations. With more aggressive commitments to research on improving energy crops and productivity, the biomass potential could be much greater. Energy-crop yield is a critical factor in estimating how much land will be needed for large-scale biofuel production,and this factor can be influenced significantly by biotechnology and systems biology strategies used in modern plant breeding and biomass processing.

Many potential energy crops (e.g., switchgrass, poplar, and willow) are essentially unimproved or have been bred only recently for biomass, compared to corn and other commercial food crops that have undergone substantial improvements in yield, disease resistance, and other agronomic traits. A more complete understanding of biological systems and application of the latest biotechnological advances will accelerate the development of new biomass crops having desirable attributes. These attributes include increased yields and processability, optimal growth in specific microclimates, better pest resistance, efficient nutrient use, and greater tolerance to moisture deficits and other sources of stress. Furthermore, many biotechnological advances for growing better biomass crops will be used to improve food crops, easing the pressure on land area needed to grow food. Joint development of these biotechnological advances with other countries will help moderate the global demand for crude oil. In an idealized future scenario with greater per-acre productivity in energy, food, and fiber crops and decreased demand for transportation fuels resulting from more efficient vehicles, the United States could have sufficient land resources to produce enough biomass to meet all its transportation-fuel needs.

A BILLION-TON ANNUAL SUPPLY OF BIOMASS: SUMMARY OF POTENTIAL FOREST AND AGRICULTURAL RESOURCES

In 2005, a study jointly supported by DOE and USDA examined whether land resources in the United States are sufficient to sustain production of over 1 billion dry tons of biomass annually, enough to displace 30% or more of the nation's current consumption of liquid transportation fuels (Perlack et al. 2005). Assuming relatively modest changes in agricultural and forestry practices, this study projects that 1.366 billion dry tons of biomass (368 million dry tons from forest and 998 million dry tons from agriculture) could be available for large-scale bioenergy and biorefinery industries by mid-21st

Century while still meeting demand for forestry products, food, and fiber (see Figure. A. Potential Biomass Resources, below). This supply of biomass would be a sevenfold increase over the 190 million dry tons of biomass per year currently used for bioenergy and bioproducts. Most of this biomass is burned for energy, with only 18 million dry tons used for biofuels (primarily corn-grain ethanol) and 6 million dry tons used for bioproducts.

Land area in the United States is about 2 billion acres, with 33% forestlands and 46% agricultural lands consisting of grasslands or pasture (26%) and croplands (20%). Of the estimated 368 million dry tons of forest biomass, 142 million dry tons already are used by the forest products industry for bioenergy and bioproducts. Several different types of biomass were considered in this study. Residues from the forest products industry include tree bark, woodchips, shavings, sawdust, miscellaneous scrap wood, and black liquor, a by-product of pulp and paper processing. Logging and site-clearing residues consist mainly of unmerchantable tree tops and small branches that currently are left onsite or burned. Forest thinning involves removing excess woody materials to reduce fire hazards and improve forest health. Fuelwood includes roundwood or logs burned for space heating or other energy uses. Urban wood residues consist primarily of municipal solid waste (MSW, e.g., organic food scraps, yard trimmings, discarded furniture, containers, and packing materials) and construction and demolition debris (see Table A. Potential Biomass Resources, this page, and Figure. B. Biomass Analysis for the Billion-Ton Study, p. 11).

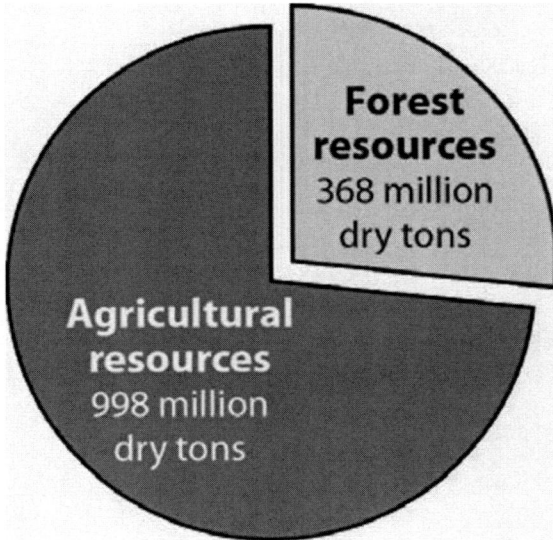

Figure. A. Potential Biomass Resources: A Total of More than 1.3 Billion Dry Tons a Year from Agricultural and Forest Resources.

Table A. Potential Biomass Resources

Biomass Resources	Million Dry Tons per Year
Forest Biomass	
Forest products industry residues	145
Logging and site-clearing residues	64
Forest thinning	60
Fuelwood	52
Urban wood residues	47
Subtotal for Forest Resources	**368**
Agricultural Biomass	
Annual crop residues	428
Perennial crops	377
Miscellaneous process residues, manure	106
Grains	87
Subtotal for Agricultural Resources	**998**
Total Biomass Resource Potential	**1366**

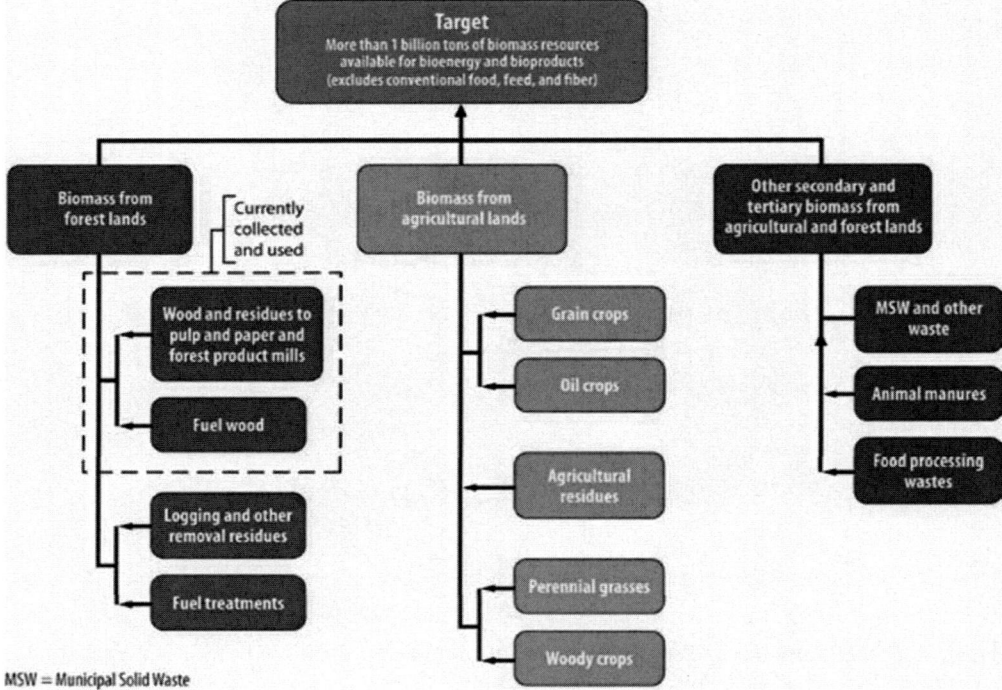

MSW = Municipal Solid Waste

Source: *Multi Year Program Plan, 2007–2012*, OBP, EERE, U.S. DOE (2005)

Figure B. Biomass Analysis for the Billion-Ton Study

Several assumptions were made to estimate potential forest biomass availability. Environmentally sensitive areas, lands without road access, and regions reserved for nontimber uses (e.g., parks and wilderness) were excluded, and equipment-recovery limitations were considered. As annual forest growth is projected to continue to exceed annual harvests, continued expansion of standing forest inventory is assumed.

Among agricultural biomass resources, annual crop residues are mostly stems and leaves (e.g., corn stover and wheat straw) from corn, wheat, soybeans, and other crops grown for food and fiber. Perennial crops considered in the study include grasses or fast-growing trees grown specifically for bioenergy. Grain primarily is corn used for ethanol production, and miscellaneous process residues include MSW and other by-products of agricultural resource processing.

A total of 448 million acres of agricultural lands, largely active and idle croplands, were included in this study; lands used permanently for pasture were not considered. Other assumptions for agricultural biomass resources include a 50% increase in corn, wheat, and small-grain yield; doubling the residue-to-grain ratio for soybeans; recovery of 75% of annual crop residues with more efficient harvesting technologies; management of all cropland with no-till methods; 55 million acres dedicated to production of perennial bioenergy crops; average biomass yield for perennial grasses and woody plants estimated at 8 dry tons per acre; conversion of all manure not used for on-farm soil improvement to biofuel; and use of all other available residues.

Agricultural Sustainability of Biomass Production

Sustainable practices for growing and harvesting biomass from dedicated crops will be essential to the success of large-scale biofuel production. Capital costs of refineries and associated facilities to convert biomass to fuels will be amortized over several decades. These capital assets will require a steady annual supply of biomass from a large proportion of surrounding land. Therefore, a thorough understanding of the conversion pathway and of biomass harvesting's long-term impacts on soil fertility is needed to ensure sustainability. Vital nutrients contained in process residues must be returned to the soil. Perennial crops expected to be used for biofuels improve soil carbon content and make highly efficient use of mineral nutrients (see sidebar, The Argument for Perennial Biomass Crops, p. 59). Additional information about the composition and population dynamics of soil microbial communities is needed, however, to determine how microbes contribute to sustaining soil productivity (see section, Ensuring Sustainability and Environmental Quality, p.68). Mixed cultivars of genetically diverse perennial energy crops may be needed to increase productivity and preserve soil quality. Because conventional annual food and fiber crops are grown as monocultures, relatively little research has been carried out on issues associated with growing mixed stands.

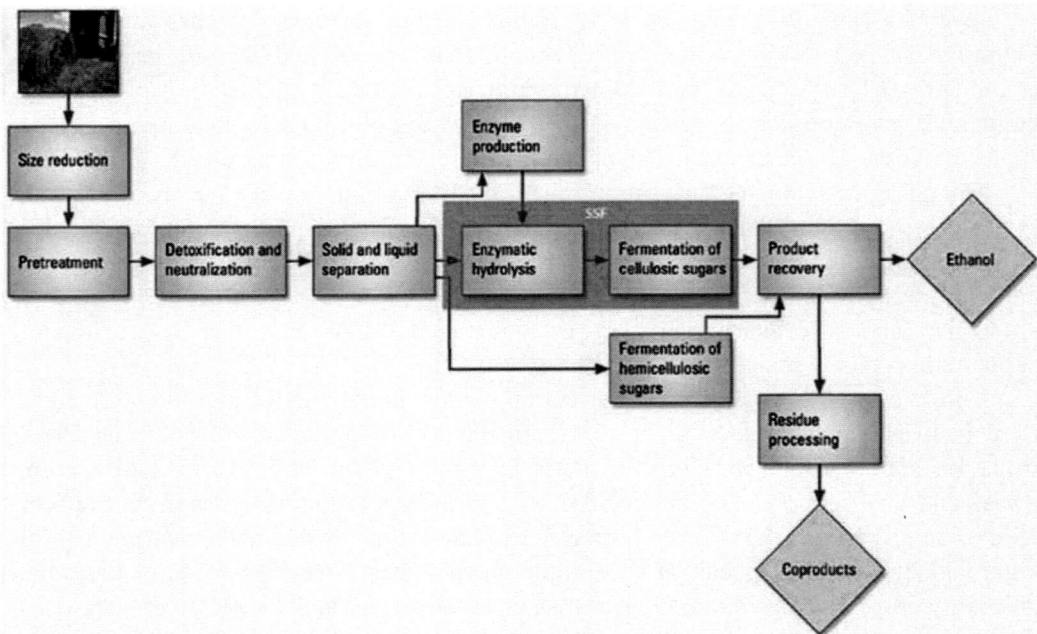

Source: Adapted from M. Himmel and J. Sheehan, National Renewable Energy Laboratory

Figure 4. Traditional Cellulosic Biomass Conversion to Ethanol Based on Concentrated Acid Pretreatment Followed by Hydrolysis and Fermentation.
Three steps in the process are (1) size reduction and thermochemical pretreatment of raw cellulosic biomass to make cellulose polymers more accessible to enzymatic breakdown and free up hemicellulosic sugars (blue boxes on left); (2) production and application of special enzyme preparations (cellulases) that hydrolyze plant cell-wall polysaccharides, producing a mixture of simple sugars (green boxes); and (3) fermentation, mediated by bacteria or yeast, to convert these sugars to ethanol and other coproducts (yellow diamonds). Recent research and development has reduced dramatically the cost of enzymes and has improved fermentation strains to enable simultaneous saccharification and fermentation (SSF, green boxes surrounded by orange), in which hydrolysis of cellulose and fermentation of glucose are combined in one step. Cellulosic biomass research is targeting these steps to simplify and increase the yield of biomass production and processing (see Figure 5. p. 15).

Today – Fuel Ethanol Production from Corn Grain (Starch Ethanol)

In 2004, 3.41 Bgal of starch ethanol fuel were produced from 1.26 billion bushels of corn—11% of all corn grain harvested in the United States. This record level of production was made possible by 81 ethanol plants located in 20 states. Completion of 16 additional plants and other expansions increased ethanol-production capacity to 4.4 Bgal by the end of 2005; additional planned capacity is on record for another 1 Bgal from 2006 to 2007 (RFA 2005). Although demand for fuel ethanol more than doubled between 2000 and 2004, ethanol satisfied less than 2% of U.S. transportation-energy demand in 2004.

In the United States, ethanol is produced in corn wet or dry mills. Corn *wet* mills fractionate the corn grain for products like germ and oil before converting the clean starch to sugars for fermentation or for such valuable food products as high-fructose corn syrup and maltodextrins.The corn fiber by-product usually is sold as animal feed. In corn *dry* mills, the grain is ground, broken into sugar monomers (saccharified), and fermented. Since the grain is not fractionated, the only by-product is the remaining solids, called distillers' dried grains with solubles, a highly nutritious protein source used in livestock feed. A bushel of corn yields about 2.5 gal ethanol from wet-mill processing and about 2.8 gal from dry grind (Bothast and Schlicher 2005). Some 75% of corn ethanol production is from dry-mill facilities and 25% from wet mills.

Tomorrow – Biorefinery Concept to Produce Fuel Ethanol from Cellulosic Biomass

Cellulosic ethanol has the potential to meet most, if not all, transportation-fuel needs. However, due to the complex structure of plant cell walls, cellulosic biomass is more difficult than starch to break down into sugars. Three key biomass polymers found in plant cell walls are cellulose, hemicellulose, and lignin (see Lignocellulosic Biomass Characteristics chapter, p. 39).These polymers are assembled into a complex nanoscale composite, not unlike reinforced concrete but with the capability to flex and grow much like a liquid crystal. The composite provides plant cell walls with strength and resistance to degradation and carries out many plant functions. Their robustness, however, makes these materials a challenge to use as substrates for biofuel production.

Traditional cellulosic biorefineries have numerous complex, costly, and energy-intensive steps that may be incompatible or reduce overall process efficiency.The current strategy for biochemical conversion of biomass to ethanol has its roots in the early days of wood chemistry. Developed in the 1930s for wartime use in Germany,it is used in Russia today. This process involves three basic steps, each element of which can be impacted by cellulosic biomass research (see Figure 4.Traditional Cellulosic Biomass Conversion to Ethanol Based on Concentrated Acid Pretreatment Followed by Hydrolysis and Fermentation, p. 14). After acquisition of suitable cellulosic biomass, biorefining begins with size reduction and thermochemical pretreatment of raw cellulosic biomass to make cellulose polymers more accessible to enzymatic breakdown and to free up hemicellulosic sugars, followed by production and application of special enzyme preparations (cellulases) for hydrolysis of plant cell-wall polysaccharides to produce simple sugars. Final steps in the process include fermentation, mediated by bacteria or yeast, to convert these sugars to ethanol and other coproducts that must be recovered from the resulting aqueous mixture. Recent research and development has reduced dramatically the cost of enzymes and has improved fermentation strains to enable simultaneous saccharification and fermentation (SSF), in which hydrolysis of cellulose and fermentation of glucose are combined in one step.

Figure 5. A Biorefinery Concept Incorporating Advanced Pretreatment and Consolidated Processing of Cellulose to Ethanol, p. 15, depicts key targets for simplifying and improving the biorefinery concept. Feedstock research seeks first to increase biomass yields and enhance biomass characteristics to enable more efficient processing. Advanced biocatalysts will augment or replace thermochemical methods to reduce the severity and increase the yield of pretreatment. More robust processes and reduction of inhibitors would allow elimination of the detoxification and separation steps. Developing modified enzymes and fermentation organisms ultimately will allow incorporation of hydrolysis enzyme production, hydrolysis, and fermentation into a single organism or a functionally versatile but stable mixed culture with multiple enzymatic capabilities. Termed consolidated bioprocessing (CBP), this could enable four components comprising steps 2 and 3 (green boxes) in Figure 4 to be combined into one, which in Figure 5 is called direct conversion of cellulose and hemicellulosic sugars. Further refinement would introduce pretreatment enzymes (ligninases and hemicellulases) into the CBP microbial systems as well, reducing to one step the entire biocatalytic processing system (pretreatment, hydrolysis, and fermentation). These process simplifications and improvements will lessen the complexity, cost, and energy intensity of the cellulosic biorefinery.

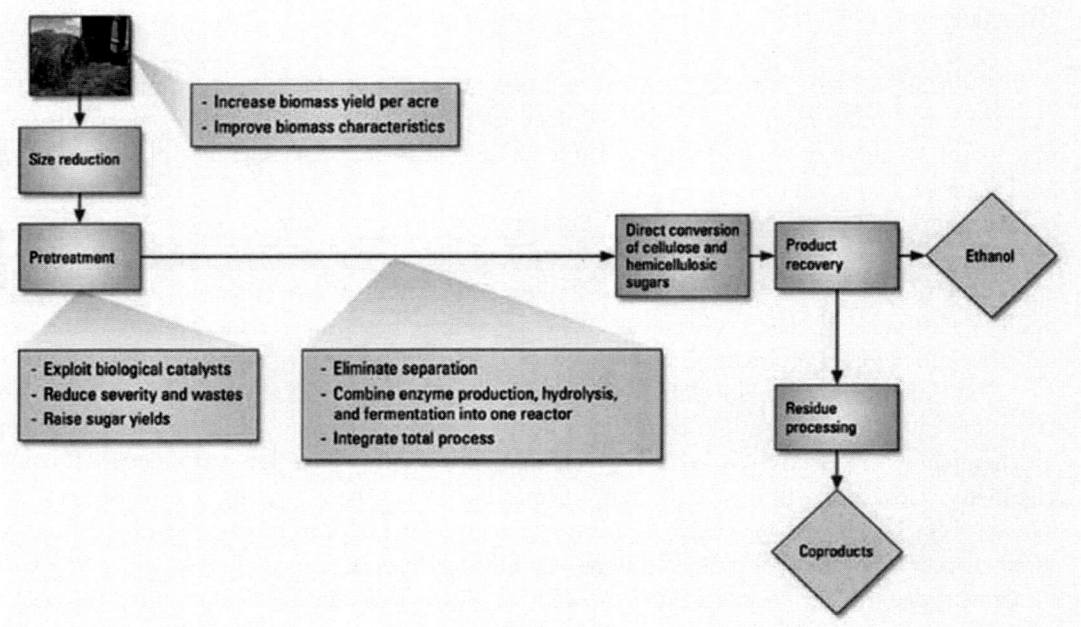

Source: Adapted from M. Himmel and J. Sheehan, National Renewable Energy Laboratory

Figure 5. A Biorefinery Concept Incorporating Advanced Pretreatment and Consolidated Processing of Cellulose to Ethanol.
The strategies discussed in this roadmap are based on first developing technologies to allow more energy-efficient and chemically benign enzymatic pretreatment. Saccharification and fermentation would be consolidated into a simple step and ultimately into a single organism or stable mixed culture (consolidated bioprocessing), thus removing multiple whole steps in converting biomass to ethanol. Also see Figure. 6. p. 16

In addition to polysaccharides that can be converted to ethanol, the lignin in plant cell walls is a complex polymer of phenylpropanoid subunits that must be separated from carbohydrates during biomass conversion.Energy-rich lignin can be burned for heat, converted to electricity consumable by other steps in the ethanol-production pathway, or gasified and converted to Fischer-Tropsch (FT) fuels (see Figure 6. Mature Biomass Refining Energy Flows: Example Scenario, p. 16, and Table A. Summary of Energy Flows in Mature Biorefinery Concept, p. 16). For more information, see Deconstructing Feedstocks to Sugars,

p. 85, and Sugar Fermentation to Ethanol, p. 119. For an overview of how genomics can be applied to developing new energy resources, see megasidebar, From Biomass to Cellulosic Ethanol, p. 26.

Ethical, Legal, and Social Issues (ELSI)

Using biomass to produce biofuels holds much promise for providing a renewable, domestically produced liquid energy source that can be a viable alternative to petroleum-based fuels. Biofuel R&D, therefore, aims to achieve more than just scientific and technological advances per se. It is conducted to accomplish important societal needs, with the broader goals of bolstering national energy security, economic growth, and the environment. Analyzing and assessing the societal implications of, and responses to, this research likewise should continue to be framed within the context of social systems and not simply in terms of technological advances and their efficacy (see sidebar, Ethical, Legal, and Social Issues for Widespread Development of Cellulosic Biofuels, this page).

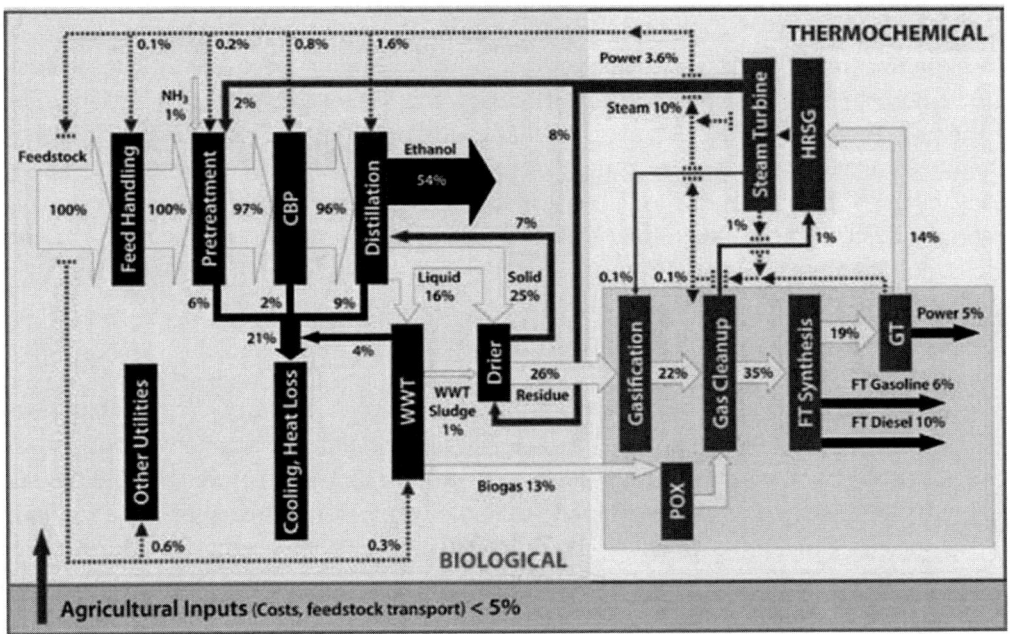

[Source: Adapted from L. Lynd et al., "Envisioning Mature Biomass Refineries," presented at First International Biorefinery Symposium, Washington, D.C. (July 20, 2005).]

Figure 6. Mature Biomass Refining Energy Flows: Example Scenario. A mature integrated cellulosic biomass biorefinery encompasses biological and thermochemical processes, demonstrating the efficiencies possible with a fully integrated design. This scenario incorporates the consolidated bioprocessing (CBP) concept, in which all biological processes are incorporated into a single microbe or microbial community. Energy derived from feedstocks is chemically and physically partitioned to ethanol and other products. Dotted arrows from above indicate energy inputs needed to run machinery. The thermochemical portion releases energy that can be used, for example, to sustain necessary temperatures, both heating and cooling, and to power pumps and other ancillary equipment. Table A is a summary of energy flows in this biorefinery concept.

Table A. Summary of Energy Flows in Mature Biorefinery Concept

Products
• 54% ethanol
• 5% power (electricity)
• 10% diesel
• 6% gasoline
Production Inputs
• 21% captured for process energy or lost
• <5% agricultural inputs (e.g., farming costs, feedstock transport)

EERE OBP PLATFORM FOR INTEGRATED BIOREFINERIES

The Department of Energy's strategic plan identifies its energy goal: "To protect our national and economic security by promoting a diverse supply and delivery of reliable, affordable, and environmentally sound energy."One of several strategies identified to achieve this goal is to "research renewable energy technologies—wind, hydropower, biomass, solar, and geothermal—and work with the private sector in developing these domestic resources."

The department's Office of Energy Efficiency and Renewable Energy (EERE) Office of the Biomass Program (OBP) elaborates on that goal: "Improve energy security by developing technologies that foster a diverse supply of reliable, affordable, and environmentally sound energy by providing for reliable delivery of energy, guarding against energy emergencies, exploring advanced technologies that make a fundamental improvement in our mix of energy options, and improving energy efficiency."

Major outcomes sought include the following.

- By 2012, complete technology development necessary to enable startup demonstration of a biorefinery producing fuels, chemicals, and power, possibly at an existing or new corn dry mill modified to process corn stover through a side stream.
- By 2012 (based on AEI), complete technology integration to demonstrate a minimum sugar selling price of $.064/lb, resulting in a minimum ethanol selling price of $1.07/gal. Ethanol would be produced from agricultural residues or dedicated perennial energy crops.
- By 2030, help enable the production of 60 billion gallons of ethanol per year in the United States. A report elaborating on this goal will be released soon.

The Biomass Program also is aligned with recommendations in the May 2001 NEP to expand the use of biomass for wide-ranging energy applications. NEP outlines a long-term strategy for developing and using leading-edge technology within the context of an integrated national energy, environmental, and economic policy.

> **ETHICAL, LEGAL, AND SOCIAL ISSUES FOR WIDESPREAD DEVELOPMENT OF CELLULOSIC BIOFUELS**
>
> Societal questions, concerns, and implications clearly may vary according to the evolutionary stage of biofuel development. Acceptance and support from diverse communities will be needed. Further, societal and technological interactions at earlier phases of research, development, demonstration, deployment, and decommissioning (RDDD&D) will affect interactions at later phases. Within the context of social systems, three overarching questions emerge.
>
> - What are the possible long-term implications of biofuel development and deployment for social institutions and systems if the strategy "works" as anticipated and if it does not?
> - How are individuals, organizations, and institutions likely to respond over time to this development and the changes integral to its deployment?
> - What actions or interventions (e.g., regulations) associated with biofuel development and its use and deployment will probably or should be taken at local, regional, and national levels to promote socially determined benefits and to avoid, minimize, or mitigate any adverse impacts?
>
> Broad topics raised at the workshop included the following:
>
> - Sustainability of the total integrated cycle.
> - Competing interests for land use.
> - Creation and use of genetically modified plants. Who creates and uses them, who decides based on what criteria, and how might or should they be regulated?
> - Creation and use of genetically modified microbial organisms in a controlled industrial setting.
> - Individuals and groups that have the authority to promote or inhibit R&D, demonstration, and use.
> - Groups most likely to be affected (positively or negatively) by biofuels at all evolutionary stages of RDDD&D on the local, national, and global levels.

The program's overarching strategic goal is to develop biorefinery-related technologies to the point that they are cost-and performance-competitive and are used by the nation's transportation, energy, chemical, and power industries to meet their market objectives. The nation will benefit by expanding clean, sustainable energy supplies while also improving its energy infrastructure and reducing GHGs and dependence on foreign oil. This goal is in alignment with DOE and EERE strategic goals as shown in Figure 7. DOE Energy Efficiency and Renewable Energy Strategic Goals as They Relate to Development of Biofuels, p. 18.

Planning documents of EERE's OBP describe advances the program seeks for four critical objectives: (1) Alter feedstocks for greater yield and for converting larger portions of raw biomass feedstocks to fuel ethanol and other chemicals; (2) decrease costs and improve enzyme activities that convert complex biomass polymers into fermentable sugars; (3) develop microbes that can efficiently convert all 5-and 6-carbon sugars released from the breakdown of complex biomass polymers; and (4) consolidate all saccharification and fermentation capabilities into a single microbe or mixed, stable culture

A commercial industry based on cellulosic biomass bioconversion to ethanol does not yet exist in the United States, but several precommercial facilities are in development. The Canadian company, Iogen Corporation, a leading producer of cellulase enzymes, operates the largest demonstration biomass-to-ethanol facility, with a capacity of 1 Mgal/year; production of cellulosic ethanol from wheat straw began at Iogen in April 2004. OBP has issued a solicitation for demonstration of cellulosic biorefineries (U.S.Congress 2005, Section 932) as part of the presidential Biofuels Initiative.

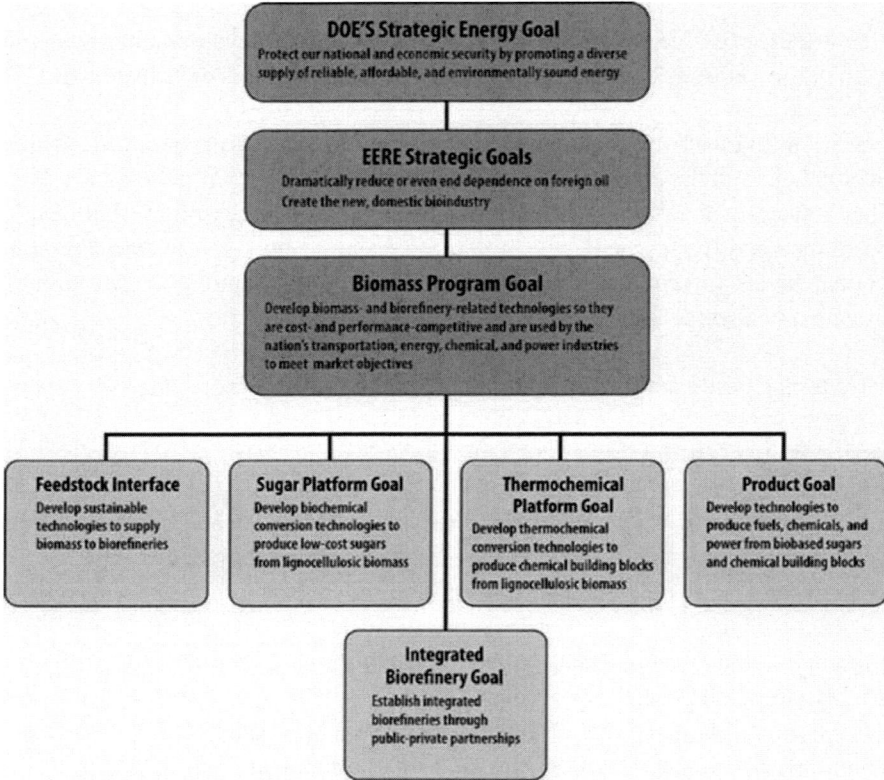

Source: *Multi Year Program Plan 2007–2012*, OBP, EERE, U.S. DOE (2005)

Figure 7. DOE Energy Efficiency and Renewable Energy Strategic Goals as They Relate to Development of Biofuels.

DOE OFFICE OF SCIENCE PROGRAMS

The DOE Office of Science (SC) plays key roles in U.S. research, including the contribution of essential scientific foundations to DOE's national energy, environment, and economic security missions (see Figure 8. DOE Office of Science Programs and Goals as They Relate to Development of Biofuels, p. 20). Other roles are to build and operate major research facilities with open access by the scientific community and to support core capabilities, theories, experiments, and simulations at the extreme limits of science. An SC goal for the Office of Biological and Environmental Research (OBER) is to "harness the power of our living world and provide the biological and environmental discoveries necessary to clean and protect our environment and offer new energy alternatives." SC's goal for its Office of Advanced Scientific Computing Research (OASCR) is "to deliver computing for the frontiers of science" (U.S. DOE 2004).

To address these priorities, OBER and OASCR are sponsoring the Genomics:Genomes to Life (GTL) program. Established in 2002, GTL uses genome data as the underpinnings for investigations of biological systems with capabilities relevant to DOE energy and environmental missions. The GTL scientific program was developed with input from hundreds of scientists from universities, private industry, other federal agencies, and DOE national laboratories. Providing solutions to major national problems, biology and industrial biotechnology will serve as an engine for economic competitiveness in the 21st Century. DOE missions in energy security are grand challenges for a new generation of biological research. SC will work with EERE to bring together biology, computing, physical sciences, bioprocess engineering, and technology development for the focused and large-scale research effort needed—from scientific investigations to commercialization in the marketplace. Research conducted by the biofuel R&D community using SC programs and research facilities will play a critical role in developing future biorefineries and ensuring the success of EERE OBP's plans.

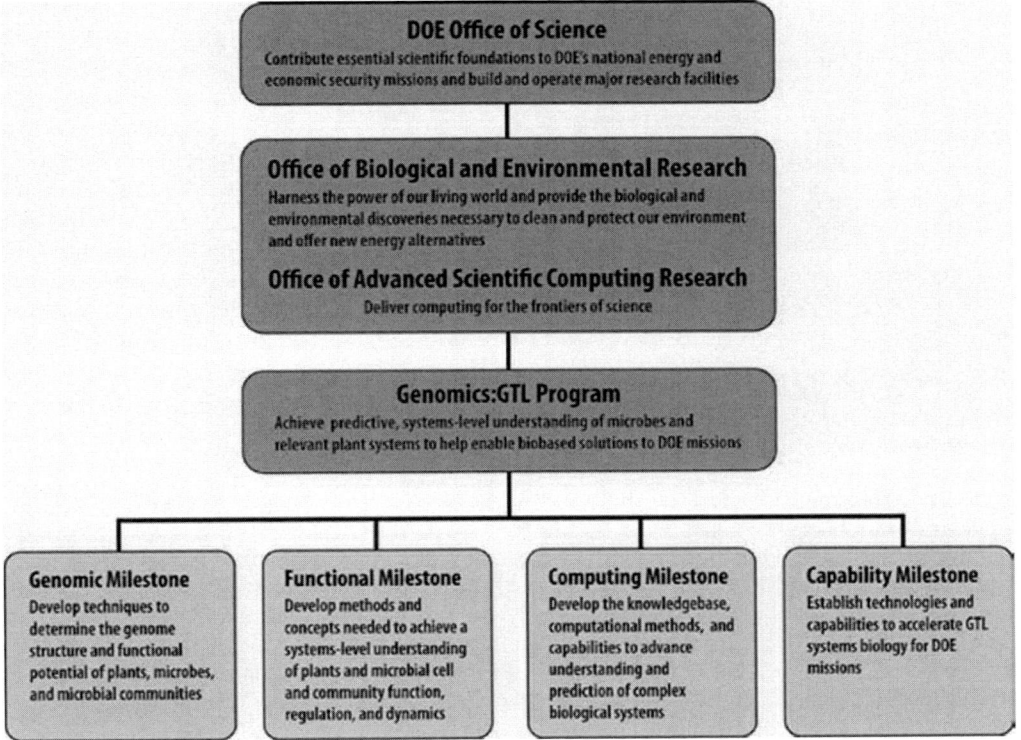

Figure 8. DOE Office of Science Programs and Goals as They Relate to Development of Biofuels. [Derived from Office of Science Strategic Plan and Genomics:GTL Roadmap]

The nation's investment in genomics over the past 20 years now enables rapid determination and subsequent interpretation of the complete DNA sequence of any organism. Because it reveals the blueprint for life, genomics is the launching point for an integrated and mechanistic systems understanding of biological function. It is a new link between biological research and biotechnology.

GTL's goal is simple in concept but challenging in practice—to reveal how the static information in genome sequences drives the intricate and dynamic processes of life. Through predictive models of these life processes and supporting research infrastructure, GTL seeks to harness the capabilities of living systems. GTL will study critical properties and processes on four systems levels—molecular, cellular, organismal, and community—each requiring advances in fundamental capabilities and concepts. These same concepts and capabilities can be employed by bioprocess engineers to bring new technologies rapidly to the marketplace.

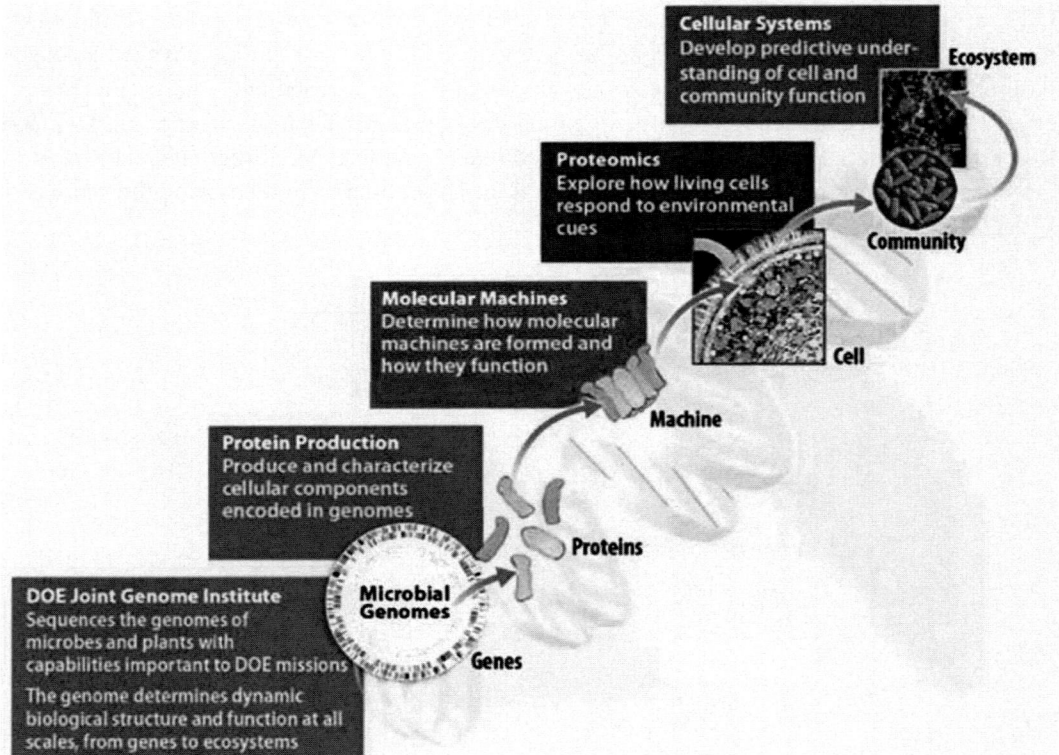

Figure 9. Understanding Biological Capabilities at All Scales Needed to Support Systems Biology Investigations of Cellulosic Biomass.
Capabilities are needed to bring together the biological, physical, computational, and engineering sciences to create a new infrastructure for biology and the industrial biotechnology in the 21st Century. This figure depicts the focus of GTL on building an integrated body of knowledge about behavior, from genomic interactions through ecosystem changes. Simultaneously studying multiple systems related to various aspects of the biofuel problem is powerfully synergistic because enduring biological themes are shared and general principles governing response, structure, and function apply throughout.
Accumulating data as they are produced, the GTL Knowledgebase and GTL computational environment will interactively link the capabilities and research efforts, allowing this information to be integrated into a predictive understanding. DOE's technology programs can work with industry to apply such capabilities and knowledge to a new generation of processes, products, and industries.

Achieving GTL goals requires major advances in the ability to measure the phenomenology of living systems and to incorporate their operating principles into computational models and simulations that accurately represent biological systems. To make GTL science and biological research more broadly tractable, timely, and affordable, GTL will develop comprehensive suites of capabilities delivering economies of scale and enhanced performance (see Figure. 9. Understanding Biological Capabilities at All Scales Needed to Support Systems Biology Investigations of Cellulosic Biomass, this page). In vertically integrated bioenergy research centers, these capabilities will include the advanced technologies and state-of-the-art computing needed to better understand genomic potential, cellular responses, regulation, and behaviors of biological systems. Computing and information technologies are central to the GTL program's success because they will allow scientists to surmount the barrier of complexity now preventing them from deducing biological function directly from genome sequence. GTL will create an integrated computational environment that will link experimental data of unprecedented quantity and dimensionality with theory, modeling, and simulation to uncover fundamental biological principles and to develop and test systems theory for biology.

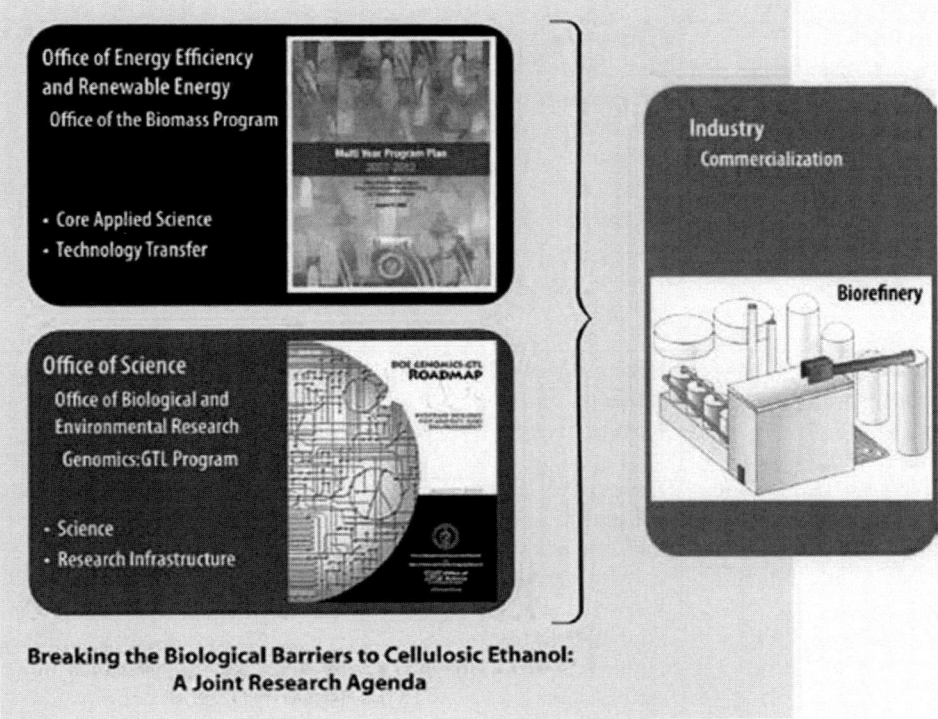

Figure 10. Creating a Common Research Agenda. The EERE Office of the Bio-mass Program's *Multi Year Program Plan 2007–2012* contains a roadmap for biofuel development that identifies technological barriers to achieving goals defined in Figure 7, p. 18. These challenges include the need for new feedstocks, their deconstruction to fermentable sugars, and fermentation of all sugars to ethanol. Within the DOE Office of Science, OBER and OASCR's roadmap for the Genomics:GTL program outlines scientific goals, technologies, computing needs, and a resource strategy to achieve the GTL goal of a predictive understanding of biological systems. This document is a roadmap that links the two plans.

BIOMASS TO BIOFUELS WORKSHOP: CREATING A COMMON RESEARCH AGENDA TO OVERCOME TECHNOLOGY BARRIERS

A product of the Biomass to Biofuels Workshop, this roadmap analyzes barriers to achieving OBP goals (as described herein) and determines fundamental research and capabilities (as described in the GTL Road-map) that could both accelerate progress in removing barriers and allow a more robust set of endpoints (see Figure. 10. Creating a Common Research Agenda, this page). Relating high-level topical areas and their goals to key scientific milestones identified by workshop participants could help achieve progress toward OBP goals in collaboration with SC (see Table 2. Overcoming Barriers to Cellulosic Ethanol: OBP Biological and Technological Research Milestones, p. 23).

Table 2. Overcoming Barriers to Cellulosic Ethanol: OBP Biological and Technological Research Milestones

Office of the Biomass Program (OBP) Barrier Topic	Technology Goals	Science Research Milestones
Feedstocks Develop sustainable technologies to supply biomass to biorefineries	**Better compositions and structures for sugars production** Domestication: Yield, tolerance Better agronomics Sustainability	Cell-wall architecture and makeup relative to processability Genome sequence for energy crops Domestication traits: Yield, tolerance Cell-wall genes, principles, factors New model systems to apply modern biology tools Soil microbial community dynamics for determining sustainability
Feedstock Deconstruction to Sugars Develop biochemical conversion technologies to produce low-cost sugars from lignocellulosic biomass	**Pretreatment Enzymes** Reduced severity Reduced waste Higher sugar yields Reduced inhibitors Reduction in nonfermentable sugars **Enzyme Hydrolysis to Sugars** Higher specific activity Higher thermal tolerance Reduced product inhibition Broader substrate range Cellulases and cellulosomes	Cell-wall structure with respect to degradation Modification of the chemical backbone of hemicellulose materials to reduce the number of nonfermentable and derivatized enzymes Cell-wall component response to pretreatments Principles for improved cellulases, ligninases, hemicellulases Understanding of cellulosome regulation and activity Action of enzymes on insoluble substrates (fundamental limits) Fungal enzyme-production factors Nonspecific adsorption of enzymes Origin of inhibitors

Table 2. (Continued)

Office of the Biomass Program (OBP) Barrier Topic	Technology Goals	Science Research Milestones
Sugar Fermentation to Ethanol Develop technologies to produce fuels, chemicals, and power from biobased sugars and chemical building blocks	**Cofermentation of Sugars** C-5 and C-6 sugar microbes Robust process tolerance Resistance to inhibitors Marketable by-products	Full microbial system regulation and control Rapid tools for manipulation of novel microbes Utilization of all sugars Sugar transporters Response of microorganisms to stress New microbial platforms Microbial community dynamics and control
Consolidated Processing Reduce process steps and complexity by integrating multiple processes in single reactors	**Enzyme Production, Hydrolysis, and Cofermentation Combined in One Reactor** Production of hydrolytic enzymes Fermentation of needed products (ethanol) Process tolerance Stable integrated traits All processes combined in a single microbe or stable culture	Fundamentals of microbial cellulose utilization Understanding and control of regulatory processes Engineering of multigenic traits Process tolerance Improved gene-transfer systems for microbial engineering Understanding of transgenic hydrolysis and fermentation enzymes and pathways

The workshop was organized under the following topical areas: Feedstocks for Biofuels (p. 57); Deconstructing Feedstocks to Sugars (p. 85); Sugar Fermentation to Ethanol (p. 119); and Crosscutting 21st Century Science, Technology, and Infrastructure for a New Generation of Biofuel Research (p. 155). A critical topic discussed in several workshop groups was Lignocellulosic Biomass Characteristics (p. 39). These five topics and plans would tie the two offices' roadmaps together and also serve as a key driver for implementing the combined roadmaps in pursuit of a high-level national goal: Create a viable cellulosic-biofuel industry as an alternative to oil for transportation. These topics and their relationships are discussed in subsequent chapters outlining technical strategy and detailed research plans developed in the workshop.

CITED REFERENCES

ACI. (2006). *American Competitiveness Initiative: Leading the World in Innovation,* Domestic Policy Council, Office of Science and Technology Policy (www.ostp.gov/html/ACIBooklet.pdf).

AEI. (2006). *Advanced Energy Initiative,* The White House, National Economic Council (www.whitehouse.gov/stateoftheunion/2006/energy/ energy_booklet.pdf).

Ag Energy Working Group. (2004). *25 by 25: Agriculture's Role in Ensuring U.S. Energy Independence* (www.bio.org/ind/25x25.pdf).

Bothast, R. J. & Schlicher, M. A. (2005). "Biotechnological Processes for Conversion of Corn into Ethanol," *Appl. Microbiol. Biotechnol.* 67, 19–25.

Brinkman, N., et al. (2005). *"Well-to-Wheels Analysis of Advanced Fuel/ Vehicle Systems—A North American Study of Energy Use, Greenhouse Gas Emissions, and Criteria Pollutant Emissions,"* General Motors Corp. and Argonne National Laboratory, www.transportation.anl.gov/pdfs/ TA/339.pdf; www.anl.gov/Media Center/News/2005/news050823 .html; www.biorefineryworkshop.com/presentations/ Wang.pdf

Bush George, W. (2006). "State of the Union Address," Washington, D.C., Jan. 31, 2006 (www.whitehouse.gov/stateoftheunion/2006/).

BTAC. (2002). *Roadmap for Biomass Technologies in the United States,* Biomass Research and Development Technical Advisory Committee (www. biomass.govtools.us/pdfs/FinalBiomassRoadmap.pdf).

De La Torre Ugarte, D. G., et al. 2003. *The Economic Impacts of Bioenergy Crop Production on U.S. Agriculture*, U.S. Department of Agriculture and Department of Energy, Agricultural Economic Report No. 816 (www. usda.gov/oce/reports/energy/AER816Bi.pdf).

EIA. (2006). *Annual Energy Outlook 2006 with Projections to 2030,* DOE/ EIA-0383, Energy Information Administration, U.S. Department of Energy (www.eia.doe.gov/oiaf/aeo/).

Ethanol Across America. (2005). "Issue Brief: Energy Security" (www.ethanol.org/documents/EnergySecurityIssueBrief_000.pdf).

Farrell, A. E., et al. (2006). "Ethanol Can Contribute to Energy and Environmental Goals," *Science* 311, 506–8.

GEC. (2005). "Ethanol From Biomass: America's 21st Century Transportation Fuel, Recommendations," Governors' Ethanol Coalition (www.ethanol-gec.org/coalitionstudies.htm).

Greene, N., et al. (2004). *Growing Energy: How Biofuels Can Help End America's Oil Dependence*, Natural Resources Defense Council, New York (www.nrdc.org/air/energy/biofuels/biofuels.pdf).

Hammerschlag, R. (2006). "Ethanol's Energy Return on Investment: A Survey of the Literature 1999–Present," *Environ. Sci. Technol.* 40, 1744–50.

IEA. (2004). *Biofuels for Transport: An International Perspective*, International Energy Agency (www.iea.org/textbase/nppdf/free/2004/biofuels2004.pdf).

Lovins, A. B., et al. (2005). *Winning the Oil Endgame: Innovation for Profits, Jobs, and Security*, Rocky Mountain Institute, Snowmass, Colorado.

Multi Year Program Plan 2007–2012. (2005). Office of the Biomass Program, Energy Efficiency and Renewable Energy, U.S. Department of Energy (www.eere.energy.gov/biomass/pdfs/mypp.pdf).

NCEP. (2004). *Ending the Energy Stalemate: A Bipartisan Strategy to Meet America's Energy Challenges*, National Commission on Energy Policy (www.energycommission.org/site/page.php?report=13).

NEPDG. (2001). *National Energy Policy: Reliable, Affordable, and Environmentally Sound Energy for America's Future*, National Energy Policy Development Group (www.whitehouse.gov/energy/National-Energy-Policy.pdf).

NRDC. (2006). *Ethanol: Energy Well Spent. A Survey of Studies Published since 1990*, Natural Resources Defense Council and Climate Solutions, New York City (www.nrdc.org/air/transportation/ethanol/ethanol.pdf).

Perlack, R. D., et al. (2005). *Biomass as Feedstock for a Bioenergy and Bioproducts Industry: The Technical Feasibility of a Billion-Ton Annual Supply*, DOE/ GO-102005-2135, Oak Ridge National Laboratory, Oak Ridge, Tennessee (http://feedstockreview.ornl.gov/pdf/billion_ton_vision.pdf).

Petrulis, M., Sommer, J. & Hines. F. (1993). *Ethanol Production and Employment*, Agriculture Information Bulletin No. 678, U.S. Department of Agriculture Economic Research Service.

RFA. (2005). *Homegrown for the Homeland: Ethanol Industry Outlook 2005*, Renewable Fuels Association (www.ethanolrfa.org/resource/outlook/).

U.S. Commerce Dept. (2006). "U.S. International Trade in Goods and Services," *U.S. Census Bureau U.S. Bureau of Economic Analysis News*, April 2006.

U.S. Congress. (2005). *Energy Policy Act of 2005*, Pub. L. 109–58.

U.S. Congress. (2002). *Farm Security and Rural Investment Act of 2002*, Pub. L. 107–71, Title IX, Sections 9001–10.

U.S. Congress. (2000). *Biomass Research and Development Act of 2000*, Pub. L. 106–224, Sections 301–10.

U.S. DOE. (2005). *Genomics:GTL Roadmap: Systems Biology for Energy and Environment*, U.S. Department of Energy Office of Science (doegenomestolife.org/roadmap/).

U.S. DOE. (2004). *Office of Science Strategic Plan*, U.S. Department of Energy Office of Science (www.er.doe.gov/Sub/Mission/Mission_Strategic.htm).

Wang, M.(2005). "Energy and Greenhouse Gas Emissions Impacts of Fuel Ethanol," Ethanol Open Energy Forum, Sponsored by the National Corn Growers Association, National Press Club, Washington, D.C. (www.anl. gov/Media_Center/News/2005/NCGA_Ethanol_Meeting_050823 .html).

BACKGROUND READING

Yergin, D. 1992. The Prize: The Epic Quest for Oil, Money, and Power, Simon & Schuster, New York.

U.S. Department of Energy
From BIOMASS to CELLULOSIC ETHANOL:

Cellulosic Biomass Feedstock
Plant Residues and Energy Crops

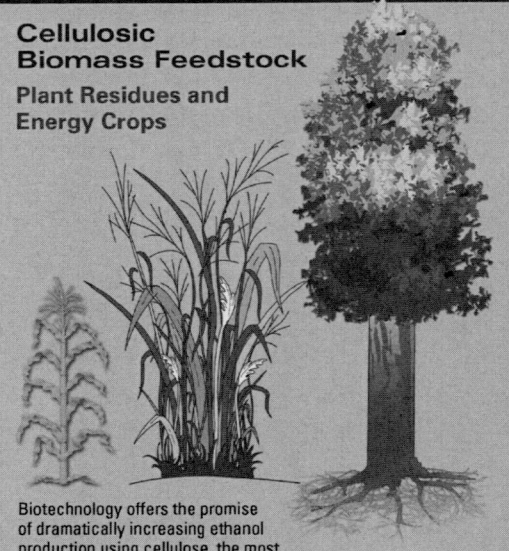

Biotechnology offers the promise of dramatically increasing ethanol production using cellulose, the most abundant biological material on earth, and other polysaccharides (hemicellulose) located in plant cell walls (see details on reverse). Residue including postharvest corn plants (stover) and timber residues could be used, as well as such specialized high-biomass "energy" crops as domesticated poplar trees and switchgrass.

Biochemical conversion of cellulosic biomass to ethanol for transportation fuel currently involves three basic steps:

▸ Pretreatments to increase the accessibility of cellulose to enzymes and solubilize hemicellulose sugars

▸ Hydrolysis with special enzyme preparations to break down cellulose to sugars

▸ Fermentation to ethanol

Making cellulosic biomass conversion to ethanol more economical and practical will require a science base for molecular redesign of numerous enzymes, biochemical pathways, and full cellular systems.

DOE GTL program contributions needed to
- Control cell-wall composition for energy production
- Develop appropriate model systems for energy crops
- Improve quantity and quality of perennial herbaceous and woody biomass crops
- Domesticate energy crops for stress tolerance
- Develop sustainable management practices

May 2007

Pretreatment
Goal: Make cellulose more accessible to enzymatic breakdown (hydrolysis) and solubilize hemicellulose sugars

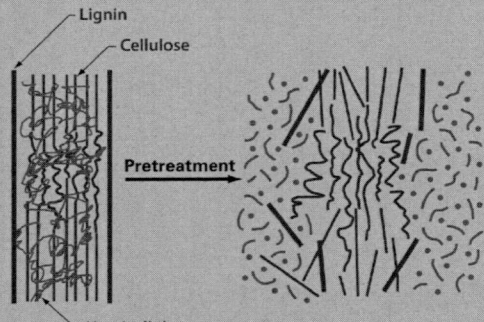

Plant cell wall

In plant cell walls (see reverse), cellulose exists within a matrix of other polymers, primarily hemicellulose and lignin. Pretreatment of biomass with heat, enzymes, or acids removes these polymers from the cellulose core before hydrolysis.

Pretreatment, one of the more expensive processing steps, has great potential for improvement through R&D.

[Figure adapted from N. Mosier et al. 2005. "Features of Promising Technologies for Pretreatment of Lignocellulosic Biomass," Bioresource Technology **96**(3), 673–86.]

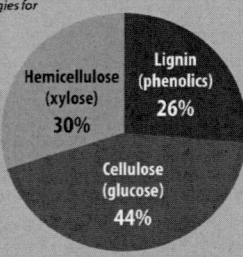

Composition of Biomass (lignocellulose)
- Hemicellulose (xylose) 30%
- Lignin (phenolics) 26%
- Cellulose (glucose) 44%

DOE GTL program contributions needed to
- Optimize and exploit biological catalysts
- Reduce thermochemical treatments and waste
- Increase simple sugar yields and concentration

All recommendations for contributions from the DOE Office of Science's (SC) Genomics:GTL program (formerly Genomes to Life) originate from a December 2005 workshop sponsored by SC and the DOE Office of Energy Efficiency and Renewable Energy. The workshop report and this flyer are available at www.genomicsgtl.energy.gov/biofuels/.

Introduction

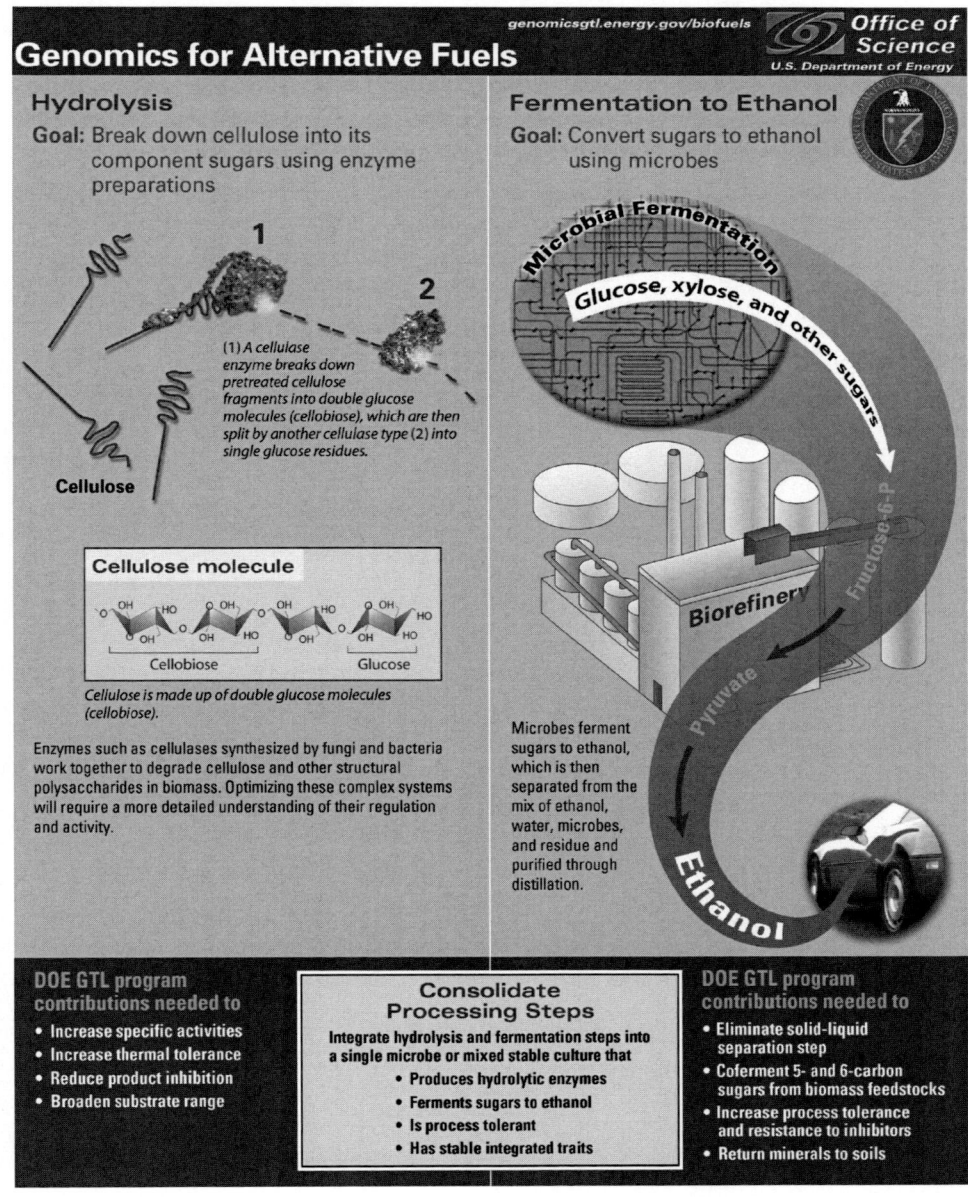

End Notes

*Gasoline and diesel constituted 98% of domestic transportation motor fuels in 2004, with ethanol from corn grain supplying most of the remaining 2%. Annual gasoline consumption in 2004 was about 139 billion gallons, and 3.4 billion gallons of ethanol were used primarily as a fuel extender to boost gasoline octane levels and improve vehicle emissions.

** Cellulosic biomass, also called lignocellulosic biomass, is a complex composite material consisting primarily of cellulose and hemicellulose (structural carbohydrates) bonded to lignin in plant cell walls. For simplification, we use the term cellulosic biomass.

*** In 2004, 11% of the U.S. corn harvest yielded 3.4 billion gallons of ethanol (NRDC 2006), roughly 1.7% of the 2004 fuel demand. Thus if all corn grain now grown in the United States were converted to ethanol, it would satisfy about 15% of current transportation needs.

In: Biological Barriers to Cellulosic Ethanol
Editor: Ernest V. Burkheisser

ISBN: 978-1-60692-203-3
© 2010 Nova Science Publishers, Inc.

Chapter 2

TECHNICAL STRATEGY: DEVELOPMENT OF A VIABLE CELLULOSIC BIOMASS TO BIOFUEL INDUSTRY

United States Department of Energy

Innovative energy crops—plants specifically designed for industrial processing to biofuels—can be developed concurrently with new biorefinery treatment and conversion processes. Recent advances in science and capabilities, especially those from the nascent discipline of systems biology, promise to accelerate and enhance this paradigm. Resulting technologies will allow the fusion of agriculture, industrial biotechnology, and energy value chains to enable an efficient and economically viable industry for conversion of plant biomass to liquid fuels. Displacing up to 30% of the nation's current transportation fuel by 2030 will require significant policy support and technical advancement over the next 5 to 15 years. Research and technology development described in this roadmap will occur in three phases to enable industry to meet the 2025 and 2030 goals (see Figure. 1. Phased Development of Bioenergy Systems, p. 32). In the Research Phase (this page), application of genome-based systems biology will provide the basis of knowledge, concepts, and tools for concerted research and deployment of technology modules in the Technology Deployment Phase (p. 32). In the Systems Integration Phase (p. 34), both fundamental and applied research and technology development will support multiple bioenergy systems through concurrent development of crops and biorefinery processes for various U.S. agroecosystems. Drivers for each phase and the research targets and goals for Feedstocks, Feedstock Deconstruction, and Fermentation to Ethanol and Recovery are outlined in Table 1. Technical Strategy Timeline, p. 33.

RESEARCH PHASE (WITHIN 5 YEARS)

Optimizing cellulose processing by refining biomass pretreatment and converting crop residues, first-generation energy crops, and other sources to liquid fuels will be the near-term focus. This will entail reducing cost, enhancing feedstock deconstruction, improving enzyme action and stability, and developing fermentation technologies to more efficiently use sugars resulting from cellulose breakdown. One goal is to decrease industrial risk from a first-of-a-kind technology, allowing more rapid deployment of improved methods.

Feedstock Use and Optimization

A range of plant materials (e.g., corn stover and hard woods) with widely varying physical and chemical characteristics could be made available as feedstocks for conversion to ethanol in biorefineries. These legacy feedstocks are expected to satisfy one-fourth to one-third of the nation's anticipated transportation biofuel needs. To achieve higher production goals, new energy crops with greater yield per acre and improved processability are needed. Advanced genome-based capabilities will help determine how soil microbial communities function and how much carbon from crop residues and dedicated energy crops, as well as other nutrients, is needed to sustain soil ecosystem function and productivity.

To establish a new generation of plants as energy crops and develop stable agroecosystems, biological and chemical tools are needed to provide detailed understanding of plant cell walls, their roles in plant function, and factors controlling recalcitrance and optimization of processes for fermenta tion of sugars. Genome-based capabilities will identify genes involved in the synthesis of cell-wall polymers and higher structures; reactions performed by the multitude of enzymes involved; design principles of cell walls; and factors controlling the amounts, composition, and structure of polymers and polymer matrices. The complex structures of plant cell walls perform numerous critical functions in the plant's growth and maintenance. Only some of these functions are now understood. Plant engineering's end goal is to use rational design for preserving critical plant functions to maximize yield and agroeconomic factors, while optimizing plant biomass makeup and structure for creating biofuels and other products. Once desirable cell-wall traits are established for energy crops, modified varieties must be domesticated for robustness and yield, and bioprocessing steps must be adapted to the superior properties of these varieties.

Plant design, bioprocess engineering, and biomass-processing strategies are intimately linked. Plants have evolved complex mechanisms for resisting assault on their structural sugars (wall polymers) from the microbial and animal kingdoms. Cell-wall polymer organization and interactions are formidable barriers to access by depolymerizing enzymes and must be deconstructed in the pretreatment step to obtain adequate rates of release and sugar yields.

Deconstruction

Understanding factors governing plant cell-wall recalcitrance to processing and deconstruction-enzyme interactions with the cell-wall matrix is critical to achieving the integrated biorefinery concept. Current technologies for biomass pretreatment (breaking down lignin and hemicelluloses and freeing crystalline cellulose) rely on thermochemical processing, which degrades the resulting biomass and results in accumulation of inhibitors that are toxic to ensuing biorefining processes. These harsh and energy- intensive pretreatments must be replaced by more benign procedures, including those that will enhance existing hydrothermal and mechanical methods and free cellulose microfibrils. Economical enzymatic procedures that would allow greater potential for integrated biorefinery strategies also will be incorporated.

Genomics and the application of such relatively new tools as proteomics, metabolomics, and imaging will aid investigators in mining burgeoning genomic databases to understand how microorganisms attack biomass and take advantage of natural enzyme diversity. Available enzymes have relatively low specific activities, so enzyme interactions, mechanisms of action, and fundamental limits must be explored. We need to determine whether low activities reflect a fundamental limit to the hydrolysis rate of certain substrates or if rates can be improved by rational design through experimentation and modeling. New classes of ligninases and hemicellulases will be identified, their mechanisms of action understood, and their performance refined to allow introduction of enzymatic pretreatment that will free cellulose microfibrils for enzymatic saccharification (breakdown to sugars).

Better methods will produce inexpensive and more robust cellulases with higher activity and longer lifetimes for breakdown of cellulose microfibrils to sugars. Mechanistic principles of cellulose-degrading enzymes will be evaluated using a range of genes and proteins found in the biosphere to understand basic design principles and allow rational redesign for enhancing properties. The diversity of cellulase and cellulosome functional schemes will be modeled and optimized for specific biomass substrates (feedstocks). The microbial cellulosome is a unique type of molecular machine that can efficiently solubilize native and pretreated lignocellulose substrates. Cellulosomes can contain the full repertoire of plant cell-wall polysaccharide-degrading enzymes, and a single cellulose-binding module presents the entire complement of enzymes to the substrate surface (see sidebar, The Cellulosome, p. 102).

Fermentation and Recovery

Use of microbes for fermentation is the most common method for converting sugars produced from biomass into liquid fuels. To develop commercially viable processes for cellulose bioconversion to ethanol, an organism is needed that uses all sugars (cofermentation of C-5 and C-6 sugars) produced from biomass saccharification at rates and in high- alcohol concentrations that match or surpass current yeast-based glucose fermentations. These capabilities, along with process-tolerant traits, involve multiple genes and pathways that are not readily resolved. Today, the capability to introduce and control multiple gene changes simultaneously in an organism is limited.

New classes of fermentation organisms and enzymes capable of metabolizing both C-5 and C-6 sugars resulting from biomass deconstruction are required to advance bioprocessing. Vast and largely untapped biochemical potential in the microbial world may be accessible through the sequencing of new microbial genomes and community "metagenomes." As first-generation organisms are being tested and improved, the focus will be on advances that allow elimination of whole steps in the conversion process. For example, thermophilic microorganisms will be examined for their ability to ferment biomass sugars at elevated temperatures, allowing development of optimal and simultaneous saccharification and cofermentation (SSCF) and thus increasing overall process efficiency. Metabolic engineering with advanced biological diagnostics will be used to develop strains with high tolerance to process stresses, inhibitors created in pretreatment, and high-alcohol concentrations. Genomic, proteomic, metabolomic, and imaging technologies, coupled with modeling and simulation, will elucidate the regulation and control of microbial metabolism and provide a predictive understanding of cell-design principles to support system engineering of integrated bioprocessing (see Figure. 1. Phased Development of Bioenergy Systems, this page, and Table 1. Technical Strategy Timeline, p. 33).

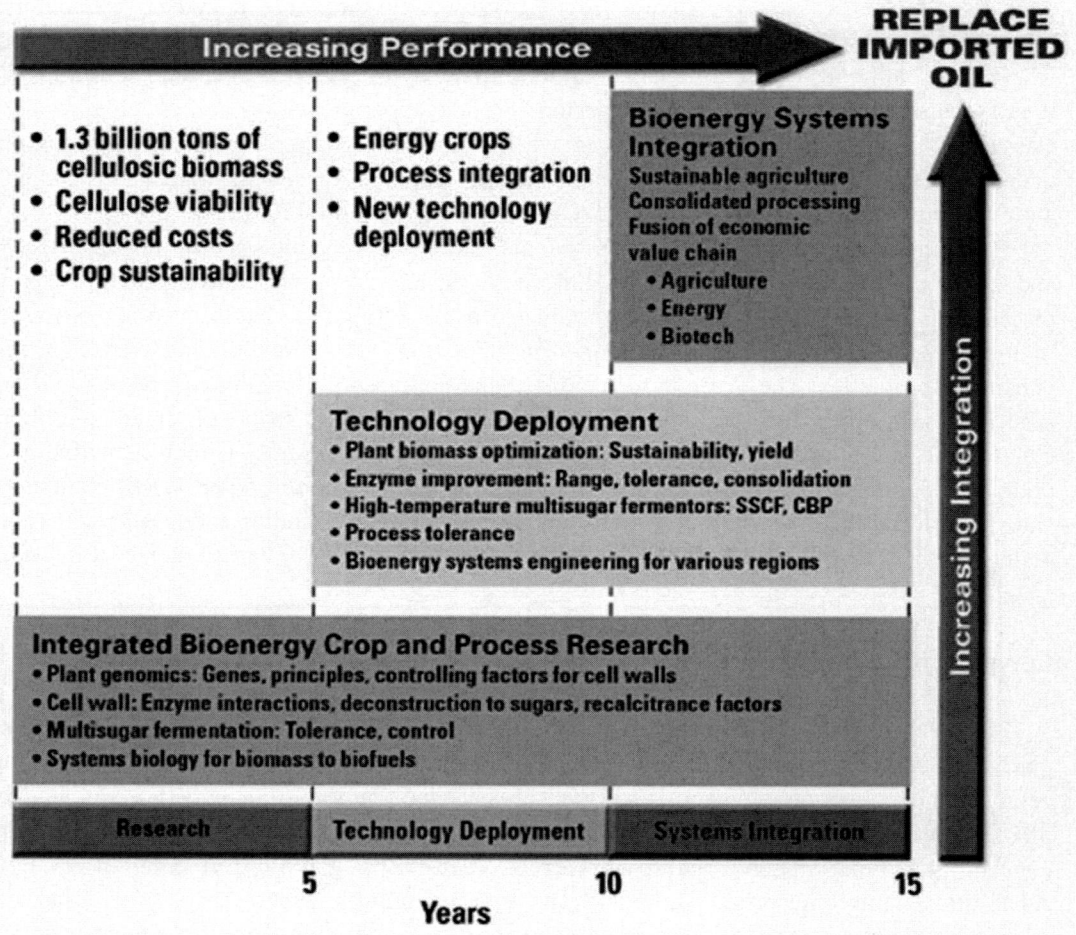

Figure 1. Phased Development of Bioenergy Systems.
Over the next 5 to 15 years, research and technology advancement will occur in three general phases. In the Research Phase, application of genome-based systems biology will provide the basis of knowledge, concepts, and tools for implementation in the Technology Deployment Phase. In the Systems Integration Phase, both fundamental and applied research and technology will support multiple bioenergy systems through the concurrrent development of crops and biorefinery processes for the various U.S. agroecosystems. Details are outlined in Table 1, p. 33.

Table 1. Technical Strategy Timeline

Research Phase (0 to 5 years)	Technology Deployment Phase (5 to 10 years)	Systems Integration Phase (10 to 15 years)
Legacy Resources and Early Energy Crops (E-Crops) Drivers Expand biomass resource base and increase utilization<ul><li>Early E-crops</li><li>Cellulose-processing viability</li><li>Cost reduction</li></ul>	**Transition: Modular Technology Deployment** Drivers<ul><li>Need for new energy crops to reach or exceed 1 billion tons of biomass</li><li>Process simplification and improvement in modules</li><li>Use of systems biology, chemistry</li><li>Bioprocess engineering</li><li>Rational systems design</li></ul>	**Integration and Consolidation** Fusion of economic value chains Drivers<ul><li>Consolidated bioenergy systems</li><li>Technologies tailored for various regions</li><li>Buildout</li></ul>
Feedstocks E-crop sustainability<ul><li>Impacts on soil ecosystems, nutrients</li><li>E-crop model development</li><li>Genes, principles, E-crop subsystem controls<ul><li>Cell-wall makeup and structure</li><li>Ties to deconstruction and fermentation</li></ul></li></ul>	**Feedstocks**<ul><li>Understand plant as system</li><li>Domesticate E-crops</li><li>Enhance sugars, minimize lignin and toxic inhibitors</li><li>Increase yield and soil sustainability</li></ul>	**End-to-End Concurrently Engineered Biofuel Systems (Biome-E Crop Processing)**<ul><li>Systems tailored to regions and fully consolidated processing</li><li>Includes E-crops with enhanced composition</li><li>Tool kits for plant engineering</li><li>Consolidated processing tied to biofuel systems<ul><li>Tailored deconstruction enzyme mix</li><li>Engineered microbial metabolic systems</li><li>Stress and process tolerance</li><li>Full system control</li></ul></li><li>Tool kits for rapid manipulation and diagnosis</li></ul>
Feedstock Deconstruction<ul><li>Reduce enzyme costs</li><li>Understand enzyme-lignocellulose interactions</li><li>Cell-wall recalcitrance</li></ul>	**Deconstruction** Deploy<ul><li>Improved enzymes (rate, specificity)<ul><li>Broadened substrate range</li><li>Reduced inhibition</li></ul></li><li>E-crop concept</li></ul>	

Research Phase (0 to 5 years)	Technology Deployment Phase (5 to 10 years)	Systems Integration Phase (10 to 15 years)
• Survey natural enzyme diversity • Establish fundamental enzyme limits • Develop ligninases and hemicellulases • Develop gene-transfer systems for cellulolytic machines and entire pathways	• Tools to diagnose and manipulate enzyme substrate interactions • Tools to design and improve enzymes	

Fermentation to Ethanol and Recovery
• Study use of all sugars, including direct cellulose utilization
• Study stress response and inhibitors
  • High-alcohol and –sugar concentrations
• Understand regulation and control
• Survey natural diversity
	Fermentation and Recovery
	Deploy
	• Cofermentation of C-5 and C-6 sugars
	• New strains (multiple)
	  • Stress tolerance
	  • High temperature
	• Tools for full regulatory control
	• Tools for rapid analysis and manipulation
	• Testing of consolidated organisms possessing celluloytic and ethanologenic properties	

TECHNOLOGY DEPLOYMENT PHASE (WITHIN 10 YEARS)

A new generation of dedicated energy crops with composition and structure better suited for breakdown to sugars for fermentation, high yield, and robustness will be essential in achieving energy security. Newly engineered deconstruction enzymes will enhance or supplant thermochemical processing to deconstruct more efficiently a broad range of biomass feedstocks. Ultimately, this will lead to enhanced energy-efficient and environmentally attractive processes for carrying out traditional pulping processes and other wood-processing techniques.

Mid-term technology will begin the consolidation of process steps. For example, improved organisms will be engineered with the ability to ferment mixed sugars, demonstrate resistance to toxic substances, and produce effective deconstruction enzymes. Systems biology and a new generation of synthetic and analytical organic chemistry will be critical for understanding these bioenergy systems and for predicting and modifying their function.

Feedstocks

Plants intended for biomass production and downstream processes involving conversion to sugars and, ultimately, ethanol will be understood and designed as a system. New breeds of energy crops will be introduced with enhanced sugar content and optimized cell-wall structures for processing, including minimization of lignin and inhibitor precursors. Plant domestication and sustainable agroecosystems based on perennials engineered to increase yield, productivity, and tolerance to such stressors as drought and salinity will reach a mature state. Multiple crops will be developed for various regional and global agroecosystems.

Deconstruction

This phase will result in deployment of improved pretreatment procedures and saccharifying enzymes with enhanced catalytic rate and substrate specificity, a broader range of applications, and reduced inhibitor production. Improved understanding of cell-wall recalcitrance and enzyme action will provide design specifications for new energy crops. Advanced high- throughput biological and chemical tools will be available to diagnose and manipulate enzyme-substrate interactions. Improved biocatalysts with desirable traits can be rationally designed for specific feedstocks and incorporated into molecular machines such as cellulosomes.

Fermentation and Recovery

New strains of industrial-processing organisms with such novel capabilities as cofermentation of C-5 and C-6 sugars and high tolerance to inhibitors, alcohol end product, and temperature will contribute to a more energy and product efficient bioprocess. Systems biology investigations will have produced a predictive understanding of cellular metabolism and regulatory controls in key fermentation microbes. This knowledge will serve as a foundation for rational development of new strains with consolidated subsets of pretreatment, hydrolysis, and fermentation capabilities. High-throughput biological and chemical tools, including computational modeling for rapid analysis and manipulation in the laboratory and in production environments, will be available.

SYSTEMS INTEGRATION PHASE (WITHIN 15 YEARS)

Bioenergy systems will include a concurrently engineered set of designer energy crops for specific agroecosystems, deconstruction and saccharification enzymes, and robust fermentation. Incorporated as multiple processes in plants or microbes, these methods will accelerate and simplify the endto-end production of fuel ethanol, enabling flexible biorefineries that can operate on a regional scale.

Integration and Consolidation

Creation of fully consolidated bioenergy systems tailored for specific regional climate and soil characteristics will allow buildout of these sustainable bioenergy economies. These systems will meld feedstocks, biomass deconstruction, bioprocess engineering, and fermentation research and development, yielding optimal two-step processes. The first step is based on consolidated feedstock traits, and the second is based on consolidated microbial traits. A concurrently engineered end-to-end biofuel system using advanced systems biology and chemical-analysis capabilities will be practicable. Toolkits for plant and agroecosystem engineering will support systems tailored to regions and consolidated processing. Companion consolidated bioprocess engineering will be tied to agroecosystems; with tailored enzyme mixes, engineered microbial metabolic systems will incorporate stress and process tolerance and permit full system control. Instrumentation available both in facilities and in the field will enable rapid diagnosis and manipulation of all critical aspects of the integrated biorefinery.

Chapter 3

SYSTEMS BIOLOGY TO OVERCOME BARRIERS TO CELLULOSIC ETHANOL

United States Department of Energy

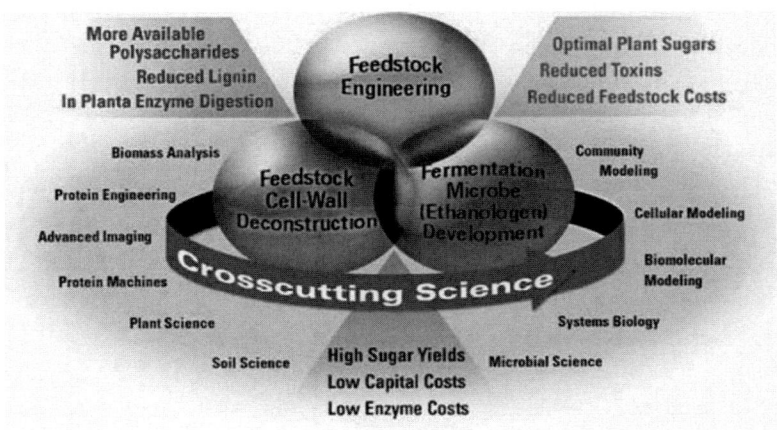

LIGNOCELLULOSIC BIOMASS CHARACTERISTICS

Makeup, Structure, and Processability

Lignocellulosic biomass has long been recognized as a potential low-cost source of mixed sugars for fermentation to fuel ethanol. Plant biomass has evolved effective mechanisms for resisting assault on its structural sugars from the microbial and animal kingdoms. This property underlies a natural recalcitrance, creating technical barriers to the cost-effective transformation of lignocellulosic biomass to fermentable sugars. Moderate yields and the resulting complex composition of sugars and inhibitory compounds lead to high processing costs. Several technologies have been developed over the past 80 years, often in wartime, that allow this conversion process to occur, yet the clear objective now is to make the process cost-competitive in today's markets.

Cell walls in lignocellulosic biomass can be converted to mixed-sugar solutions plus lignin-rich solid residues by sequential use of a range of thermochemical pretreatments and enzymatic saccharification. The low rate at which biomass is converted to sugars and the coproduction of fermentation inhibitors increase equipment size and result in high pretreatment and enzyme costs. New approaches for designing improved energy feedstocks, deconstructing plant cell walls, and transforming their polysaccharides to fermentable sugars are needed. A systematic understanding of enzyme interactions with plant cell architecture and hierarchy, as well as cellulose, hemicellulose, and lignin structure during chemical and enzymatic hydrolysis, will allow the prediction of plant-tissue response to hydrolytic attack and the creation of new systems.

Significant technology development will be needed for creation of large-scale bioenergy and biorefinery industries that can handle a billion tons made up of a variety of biomass each year. In the DOE-USDA Billion- Ton Study, corn stover and perennial crops such as switchgrass and hybrid poplar make up about half the potential 1.3 billion tons of biomass that could be available by the mid-21st Century (Perlack et al. 2005). Understanding the structure and function of these and other biomass resources will be critical to enhancing their processability.

The result of analysis and research described here will be to increase the efficiency with which the solid (substrate) interacts with large protein macromolecules (enzymes) at its surface while the surface itself is being eroded into soluble oligosaccharides [see sidebar, Image Analysis of Bioenergy Plant Cell Surfaces at the OBP Biomass Surface Characterization Lab (BSCL), p. 40]. This knowledge, combined with development of new proteins that catalyze these transformations as well as microbial systems for fermentation and consolidation, will enable the design of procedures and hardware that dramatically speed up the process, improve yield, and lower costs.

IMAGE ANALYSIS OF BIOENERGY PLANT CELL SURFACES AT THE OBP BIOMASS SURFACE CHARACTERIZATION LAB (BSCL)

Many aspects of current biomass conversion technology are becoming better understood, and a nascent biomass processing industry is emerging for some niche markets. To reach the mid- and long-term goals stated in the DOE Office of the Biomass Program's *Multi Year Program Plan: 2007-2012,* however, enhanced fundamental understanding of feedstocks and all biorefinery processes is critical. For example, detailed knowledge about plant cell-wall ultrastructure and function to formulate improved enzyme mixtures and pretreatments will reduce the cost of producing sugars. (See images, right.)

In many cases, we know how to describe biomass compositionally. That is, we can conduct chemical or spectroscopic analyses and determine the percentages of individual sugars, protein, uronic acids, and lignin. When we study biomass conversion of corn stover, hardwoods, or rice straw, for example, we are in fact working primarily with the plant's structural parts, most of which are cell wall. Therefore, more knowledge is needed about the natural organization and structure of polymers and chemicals in plant tissue that affect chemical pretreatment, enzymatic digestibility, and the generation of compounds inhibiting fermentative microorganisms used to produce the final fuel or chemical. The study of plant cell walls at the submicron or macromolecular scale is challenging. Imaging and image analysis are at the cutting edge of botany, molecular biology, biochemistry, chemistry, and material and computer sciences. Descriptions of microscopies important for ultrastructure imaging are in the Imaging Technologies section of the Crosscutting Technologies chapter, p. 163, and sidebar, Some Imaging Technologies Relevant to Feedstock Characterization, p. 163.

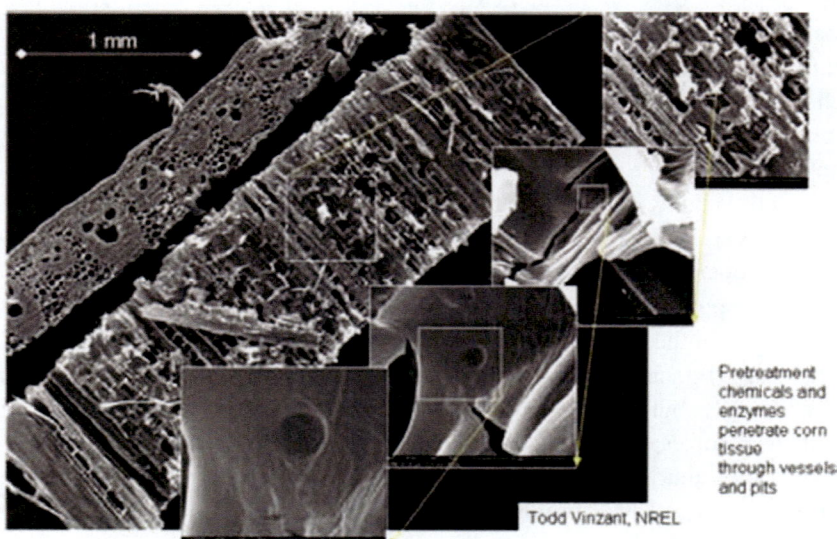

Figure A. Collage of Scanning Electron Microscopy Images Showing a Rind and Adjacent Pith Section Cut from a Field-Dried Corn-Stem Cross Section.
The rind shows a higher density of vascular elements made from thick-walled cells. The pith section (shown longitudinally) shows a greater number of thin-walled parenchyma cells. Overall, most cellulose needed for biomass conversion is located in the rind, although the pith represents most of the stem volume. Closeups of a cell-wall pit also are shown (~150,000×). These structures are thought to aid transfer of chemicals and enzymes used in processing within the biomass bulk.

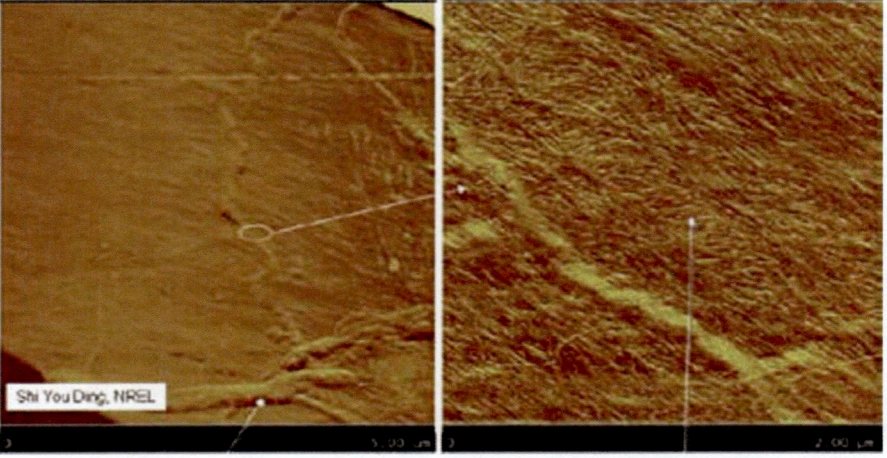

Figure B. AFM: Corn Parenchyma Cell Wall.

STRUCTURE AND ASSEMBLY OF CELL WALLS

Plant cell walls are a complex and dynamic mixture of components that perform many functions (see Figure. 1. Simplified Cell Wall and Figure. 2. Conceptual Illustration of Cell-Wall Biogenesis, p. 42; and sidebar, Understanding Biomass, pp. 53 to 55). The cell walls are intricate assemblages of celluloses, hemicelluloses (i.e., xyloglucans, arabinoxylans, and glucomannans), pectins (i.e., homogalaturonans, rhamnogalacturonan I and II, and xylogalacturonans), lignins, and proteoglycans (e.g., arabinogalactan-proteins, extensins, and proline-rich proteins). Most mass in the plant cell wall is in the form of polysaccharides (cellulose and hemicelluloses). The next most abundant polymer is lignin, which is composed predominantly of phenylpropane building blocks. Lignins perform an important role in strengthening cell walls by cross-linking polysaccharides, thus providing support to structural elements in the overall plant body. This also helps the plant resist moisture and biological attack. These properties of lignin, however, interfere with enzymatic conversion of polysaccharide components. Additionally, since lignin is not converted readily to ethanol, we must find other uses in the process if we are to maximize energy yield from biomass.

Several thousand gene products are estimated to participate in synthesis, deposition, and function of cell walls, but very few associated genes have been identified and very little is known about their corresponding enzymes. Many questions remain, for example, regarding how polysaccharides and lignin are synthesized, how wall composition is regulated, and how composition relates to cell-wall biological functions. To answer these questions, we need to discover the functions of many hundreds of enzymes, where proteins are located within cells, whether or not they are in complexes, where and when corresponding genes are expressed, and what factors and genes control expression and activities of the proteins involved. Application of new or improved biological, physical, analytical, and mathematical tools will facilitate a detailed mechanistic understanding of cell walls. That knowledge will permit optimization of various processes involved in producing biomass and converting it to fuels.

Productivity and conversion-process efficiencies can be increased by altering fundamental aspects of plant growth, development, and response to biotic and abiotic stress. Altering cell-wall composition to increase the relative amount of cellulose and decrease lignin, for example, could have significant effects (see sidebar, Optimizing Lignin Composition for More Efficient Bioethanol Production, p. 43). Eventually, a systems cell-wall model incorporating biophysical aspects with structural properties and knowledge of proteins involved in synthesis will aid in rational development of highly productive feedstock species whose cell walls are optimized for conversion.

Plants can have two types of cell walls, primary and secondary. Primary cell walls contain cellulose, which consists of hydrogen-bonded chains of thousands of β-1,4-linked glucose molecules, in addition to hemicelluloses and other materials woven into a nanoscale network with the cellulose. Cellulose in higher plants is organized into microfibrils, each measuring about 3 to 6 nm in diameter and containing up to 36 cellulose chains. Each cellulose chain is a linear collection of thousands of glucose residues. Pairs of glucose residues (cellobiose) make up the repeating unit of cellulose.

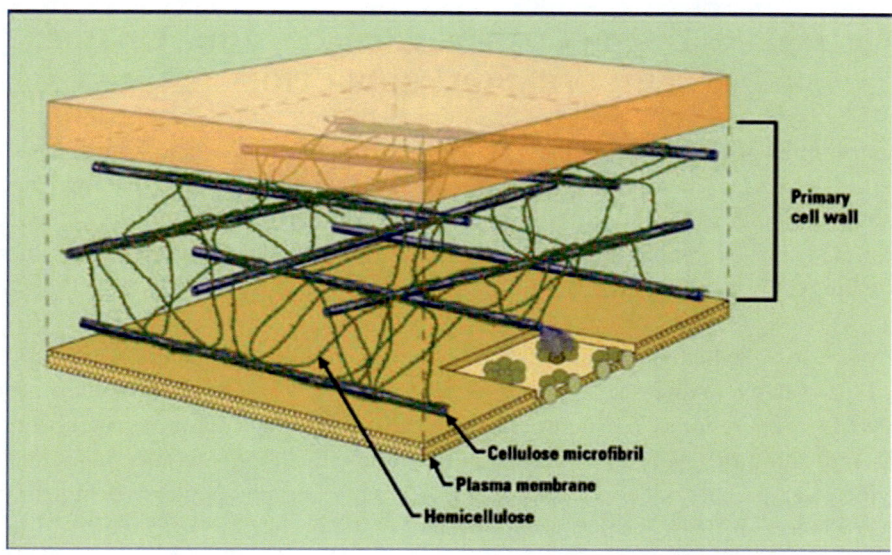

Figure 1. Simplified Cell Wall.
For more details, see sidebar, Understanding Biomass, p. 53. [Adapted with permission from C. Somerville et al., Science 306, 2206–11 (2004); © 2004 AAAS.]

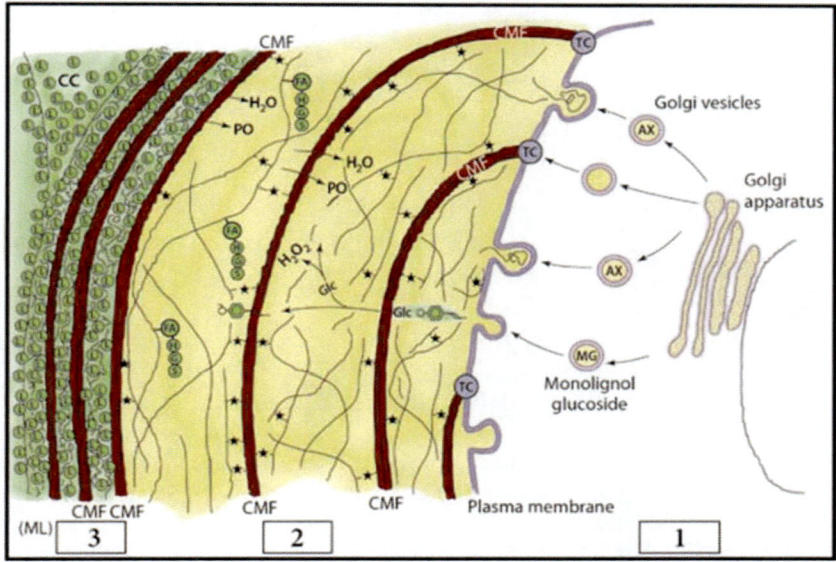

Figure 2. Conceptual Illustration of Cell-Wall Biogenesis.
The Golgi apparatus participates in hemicellulose and lignin biosynthesis. Cellulose microfibrils (CMF) are laid separately in the swollen gel of hemicelluloses. As lignin is deposited, the cell wall becomes hydrophobic. Water removal from the swollen gel, together with peroxidase (PO) and calcium, causes anisotropic shrinkage perpendicular to the CMFs. This shrinkage drives further oligolignol polymerization by orienting the lignin aromatic ring parallel to the cell-wall surface. [Adapted from N. Terashima et al., "Comprehensive Model of the Lignified Plant Cell Wall," pp. 247–70 in *Forage Cell Wall Structure and Digestibility*, ed. H. G. Jung et al., American Society of Agronomy, Crop Science Society of America, and Soil Science Society of America (1993).]

Optimizing Lignin Composition for More Efficient Bioethanol Production

Plant lignin (guaiacyl and syringyl) interferes with the release and hydrolysis of cell-wall polysaccharides. Metabolic engineering of the lignin biosynthetic pathway has been suggested as a method for modifying lignin content in feedstocks. Studies in *Arabidopsis* demonstrated that overexpression of the enzyme ferulate 5-hydroxylase (F5H) increases lignin syringyl monomer content and abolishes the tissue specificity of its deposition (Figure. A).

To determine whether or not this enzyme has a similar regulatory role in woody plants, F5H was overexpressed in poplar trees using a cinnamate 4-hydroxylase promoter to drive F5H expression. Transgenic trees displayed enhanced lignin syringyl monomer content, indicating that F5H overexpression is a viable metabolic engineering strategy for modifying lignin biosynthesis. These high-syringyl lignin poplars demonstrated a significant increase in chemical pulping efficiency. [R. Franke et al., "Modified Lignin in Tobacco and Poplar Plants Over-Expressing the *Arabidopsis* Gene Encoding Ferulate 5-Hydrosylase," *Plant J.* 22(3), 223–34 (2000).] Similar metabolic engineering strategies hold promise for developing improved feedstocks for bioethanol production.

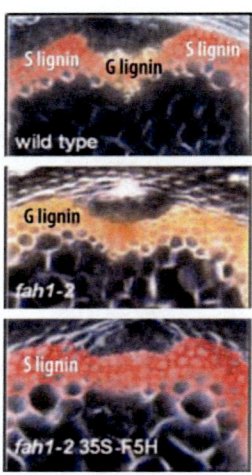

Figure A. Lignin Composition Controlled by Genetic Manipulation and Monitored via Histochemical Staining for Lignin Monomer Composition in *Arabidopsis* Stem Cross Sections. In the lignified cells of wild-type *Arabidopsis* stems, the presence of syringyl (S) or guaiacyl (G) monomers can be visualized by histochemical staining of S lignin (red) and G lignin (yellow). Histochemical staining allows a diagnosis of the effects of experiments to manipulate lignin composition. For example, eliminating one enzyme of the lignin biosynthetic pathway in an *Arabidopsis* mutant (*fah1*-2) leads to a pure G lignin, and overexpression of the same enzyme leads to a homogeneous deposition of S lignin (35S-F5H). The *fah1* gene encodes ferulate-5-hydroxylase, a cytochrome P450–dependent monooxygenase that catalyzes hydroxylation of coniferaldehyde and coniferyl alcohol in the pathway leading to syringyl lignin. [Figures published in C. Chapple et al., "Lignin Monomer Composition is Determined by the Expression of a Cytochrome P450–Dependent Monooxygenase in *Arabidopsis*," *Proc. Natl. Acad. Sci. USA* 95, 6619–23 (1998); ©1998 National Academy of Sciences, U.S.A.]

> Many aspects of lignin biosynthesis remain matters of debate. Although most genes involved in the biosynthetic pathway have been cloned and functions assigned, mechanisms that regulate the pathway still are largely unknown, as is its relationship with other cell-wall biochemical pathways and plant development. Topics to be studied include regulation of lignin deposition and tissue specificity, identity of proteins involved in monolignol transport and polymerization, and ways in which lignin content and composition can be modified (see Figure. 4. Phenylpropanoid Pathway Leading to Lignin Biosynthesis in Plants, p. 49). Also needed is a detailed understanding of lignin-biodegradation mechanisms, including that accomplished by white rot fungi, which break down yellow lignin and leave behind crystalline white cellulose (see sidebar, White Rot Fungus, p. 93). Comprehensive explorations of lignin biosynthesis and degradation are required to maximize energy yield from biomass crops.

Like steel girders stabilizing a skyscraper's structure, the mechanical strength of the primary cell wall is due mainly to the microfibril scaffold. A microfibril's crystalline and paracrystalline (amorphous) cellulose core is surrounded by hemicellulose, a branched polymer composed of pentose (5-carbon) and, in some cases, hexose (6-carbon) sugars. In addition to cross-linking individual microfibrils, hemicelluloses in secondary cell walls also form covalent associations with lignin, a complex aromatic polymer whose structure and organization within the cell wall are not completely understood (see Figure. 3. Association of Lignin with Polysaccharides, this page). The crystallinity of cellulose and its association with hemicellulose and lignin are two key challenges that prevent the efficient breakdown of cellulose into glucose molecules that can be fermented to ethanol.

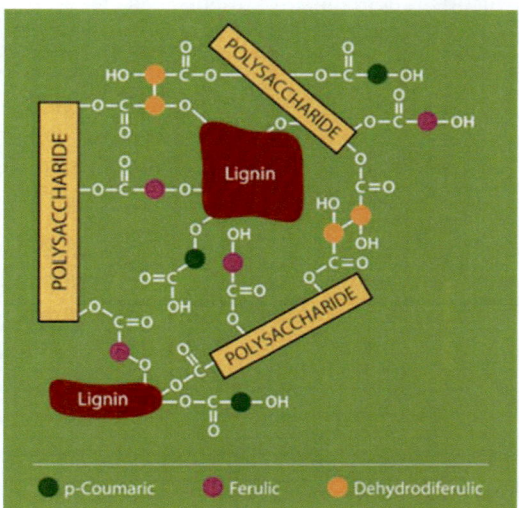

Source: Adapted from K. Iiyama, T. Lam, and B. Stone, "Covalent Cross-Links in the Cell Wall," *Plant Physiol*. 104(2), 318 (1994). Reprinted with permission of American Society of Plant Biologists, ©ASPB 1994.

Figure 3. Association of Lignin with Polysaccharides. The schematic diagram shows possible covalent cross-links between polysaccharides and lignin in cell walls. Lignin is bonded to the cellulose and hemicellulose polysaccharides and serves as a stiff- stiffening and hydrophobic agent, complicating biomass breakdown.

Many enzymes involved in cell-wall synthesis or modification are thought to be located in protein complexes. Within the plasma membrane are rosettes composed of the enzyme cellulose synthase; these protein complexes move laterally along the membrane to synthesize cellulose molecular chains (36 per rosette), which crystallize into microfibrils. Movement of the rosette molecular machine is associated with cortical microtubules that underlie the membrane, but that linkage also is poorly understood. The interaction of cellulose synthase with the cytoskeleton has an impact on cellulose fibril orientation and perhaps length. Understanding the function of these complexes and their interactions with metabolic pathways that produce sugars will be important for eventually controlling cell-wall composition. A number of cellulose-synthase genes have been cloned for a variety of plants. (See sidebar, Understanding Biomass, beginning on p. 53, for an illustrated explanation.)

FACTORS IN RECALCITRANCE OF LIGNOCELLULOSE PROCESSING TO SUGARS

Organization and interactions among polymers of the cell wall—constructed for strength and resistance to biological, physical, and chemical attack—constitute a barrier to access by depolymerizing enzymes and must be partially deconstructed in the bioconversion pretreatment step before saccharification can occur. Although various pretreatments have been developed, we still do not have a detailed understanding of fundamental physical and chemical features of lignocellulosic biomass that limit its breakdown to sugars and ultimate bioconversion efficiency. Improved methods must be developed to characterize biomass and its interaction with various chemical treatments, as well as with deconstruction and saccharification enzymes.

Natural factors believed to contribute to lignocellulosic feedstock's recalcitrance to chemicals or enzymes include plant and cell-wall architecture and molecular structure.

Plant Architecture

The organs of a plant (leaf, stem or trunk, and root) are composed of myriads of cells with different functions in the plant's economy. Each has its own particular type of cell wall whose composition is related directly to cell function [e.g., support (fibers), protection (epidermis), and transport (xylem and phloem)]. Leaf, stem, and root tissues invariably contain cells of more than one type. In tissues, individual cells are closely associated at their cell-wall interfaces to give a compact tissue structure. This structure must be disassembled by milling (comminution) to allow liquid access to cell walls.

- The waxy barrier comprising grass cuticle and tree bark impedes penetration of enzymes.
- Even milled plant stems and woody tissues limit liquid penetration by their nature.

Cell-Wall Architecture

The nanoscale composite nature of the plant cell wall restricts penetration of chemicals and enzymes to their substrates. The lignin-hemicellulose coating on the cell wall's cellulosic microfibrils affects the following:

- Conformation of cellulose and noncellulosic polysaccharides making up the microfibril limits accessibility of hydrolytic enzymes to their substrates.
- Lignin-carbohydrate complexes limit enzymatic hydrolysis of biomass polysaccharides.

Molecular Structure

Cellulose crystallinity severely restricts cellulase attacks. Cellulases must physically release individual cellulose chains from microfibril crystals for subsequent catalytic hydrolysis to sugars. Limiting factors:

- Inherent difficulty of enzymes in acting on poorly hydrated cellulose surfaces.
- Amount and composition (including heterogeneity) of lignin.
- Chemical heterogeneity and strength of covalent interactions between lignin and noncellulosic polysaccharides.
- Robustness of hydrogen bonding in cellulose microfibrils arising from extended hydrogen-bond periodicity.

In addition, all native celluloses undergo physical modifications that can inhibit saccharification as they are dehydrated in traditional methods of isolation or storage after harvest and in pretreatment processes. Characterizing the effects of conditions and storage environments would point to modifications in harvesting and storage for biomass resources. Understanding physicochemical characteristics of the cell-wall polysaccharide system would guide genomic modifications of bioenergy crops to facilitate processing and resist deleterious physical modification as much as possible. For example, structural elements of many lignocellulosic materials react to pretreatment in ways that reduce enzymatic digestibility.

- High mechanical pressure such as that from plug feeders collapses the natural vascular structure.
- Dilute-acid chemical pretreatments may permit cellulose to reanneal, leading to "hornification" of cellulose in microfibrils.
- Ambient or elevated temperatures may accelerate denaturation (e.g., the tendency of most (beta-1,4)-pentosans and hexosans to have inverse water-solubility relationships with temperature).
- Some pretreatments may permit lignin to become soluble and "plate out" on cellulose surfaces during the cool-down phase.

These "process-induced" causes of recalcitrance must be understood and overcome through process modification or biomass design.

After cellulose, hemicellulose is the next most abundant polysaccharide in native biomass feedstocks. Structural information on these polymeric substrates is necessary, and mechanistic models must be developed to identify "bottlenecks" in hemicellulose bioconversion (see sidebar, Optimizing Hemicellulose Acetylation in Cell Walls, this page).

A systematic approach to understanding these factors will promote more effective use of lignocellulosic biomass in bioconversion systems. Fortunately, we now have new biological, physical, analytical, and mathematical tools that can help in reliably identifying and quantifying the relative importance of various potentially limiting factors. We also have tools to identify and optimize facilitating factors, for example, through plant breeding.

OPTIMIZING HEMICELLULOSE ACETYLATION IN CELL WALLS

Hemicellulose Acetylation Degradation Products Are Toxic to Microbes

Acetyl side groups from hemicellulose biomass polymers are released during current pretreatment steps. These small acetyl molecules often are toxic and inhibit the microbial activity that converts sugars to ethanol. Hemicellulosic components such as xyloglucan and glucuronoarabinoxylan and pectic cell-wall components often are O-acetylated. For instance, O-acetyl groups may be present on the glucan backbone of xyloglucan or on galactose or arabinose residues of side chains. The degree of sugar-residue O-acetylation of pectins varies from 0 to 90% depending on the tissue, species, and method of preparation. The role of O-acetyl substituents in vivo is not known, but in vitro experiments suggest that one function may be their involvement in hindering enzymatic polysaccharide breakdown. O-acetyl substituents also affect polysaccharide solubility and pectin's gelation properties (Pauly and Scheller 2000).

Plant genes exhibit weak sequence similarity to putative bacterial acetyltransferase genes. Genetic tools in plants such as *Arabidopsis* will enable the identification of gene products catalyzing polysaccharide acetylation and the determination of acetylation's role in cell-wall structure and function. Such studies will provide insights into the possibility of developing biomass crop varieties with significantly reduced polysaccharide acetylation and thus improving the fermentation process.

The goal is to provide a rational basis for design of practical, effective, and economical pretreatments, including controlling the physical modification of native celluloses and related cell-wall polysaccharides during thermal and chemical treatments. Current thermochemical treatments ultimately will be replaced with more benign enzymatic treatments to the degree feasible. Necessary detailed analyses are discussed in the chapter, Deconstructing Feedstocks to Sugars, p. 85.

OPTIMIZATION OF PLANT CELL WALLS

Optimal efficiency of biofuel production depends on maximizing fuel yield from a unit of biomass and minimizing energy inputs. The plant cell wall, a complex assembly that plays a primarily, but not exclusively, structural role during plant growth and development, may be particularly amenable to the application of engineering principles in redesigning the cell wall to meet energy needs. To breed plants in which cell-wall composition is optimized for conversion efficiency, understanding how cell walls are made, how composition is regulated, and the roles of various polymers in supporting plant growth and development will be necessary. The long-term goal is to develop a systems-level understanding to facilitate rational improvement of plant cell-wall composition in dedicated energy crops. Such knowledge of plant cell walls is in a very primitive stage because of scientific and technical challenges that have impeded scientific progress. Future research on cell-wall synthesis and function requires interdisciplinary approaches ranging from genomics to synthetic carbohydrate chemistry and biophysics. Model organisms are important in facilitating advances in basic biology and in bringing the most sophisticated biological tools to the problem. Several new plant models closely related to species selected for energy crops are advocated. A powerful first step is to obtain comprehensive DNA sequences for these organisms.

Understanding Cell-Wall Structure and Function

Increasing the production of biofuels begins with increasing biomass productivity, either by making more cell walls or making cell walls with more carbon. In addition, changes in cell-wall composition could have major effects on the efficiency with which biomass can be converted to fuels; relative amounts of certain sugars could be increased or wall polymers could be made more amenable to enzymatic hydrolysis, thus improving the yield of sugars delivered to the biorefinery as raw feedstock.

Important questions remain about the structures of cell-wall polymers, how they are made, and their functions in plant growth and development. To optimize the amount, composition, and structure of walls for biofuel production, we must identify the genes involved in synthesis of cell-wall polymers, the design principles for cell walls, and factors that control the amounts and organization of various types of enzymes and resultant polymers. Preliminary evidence suggests that cell-wall biophysical properties important to plant growth and development may be achieved in many different ways with regard to chemical composition. Thus, cell-wall composition of energy crops probably can be altered so they are better suited for fuel production. Desirable improvements include increasing the amount of such useful polysaccharides as cellulose or certain hemicelluloses and minimizing the content of such undesirable components as lignin or acetyl groups.

Evidence indicates that photosynthetic CO_2 fixation is regulated by plants in response to demand for fixed carbon, so understanding photosynthate flux into cell-wall polymers relative to other pathways of primary carbon metabolism and storage is important. Understanding mechanisms that regulate carbon flux and synthesis of various polysaccharides may make possible the development of plants that accumulate significantly more polysaccharide per cell.

Expected significant increases in the ratio of carbon to nitrogen and mineral nutrients would have a beneficial effect on agricultural inputs (e.g., planting, fertilizing, cultivating, and harvesting), costs, and sustainability.

Progress in this area requires broad approaches to achieve a foundation of knowledge about cell-wall structure and function that will be the basis for a systems approach to predicting and controlling biomass composition. Before a systems approach can be implemented, a comprehensive understanding is needed about what reactions are performed by the many hundreds of enzymes involved in cell-wall synthesis and deposition, where and when relevant genes are expressed, and what genes control expression and activity of proteins involved in polysaccharide and lignin synthesis and modification. Indeed, one "grand challenge" in systems biology may be understanding how to engineer cell walls that meet the need of chemical biorefineries for optimized feedstocks yet still meet the plant's need for development, robustness, and maximal rates of growth.

GTL capabilities could provide extensive support for research on cell-wall synthesis, structure, and function. Sequencing support for model organisms (see below) and for identifying relevant genes in energy crops is an immediate goal. The development of populations of transgenic experimental plants with epitope-tagged proteins would greatly facilitate the determination of subcellular protein localization and the application of proteomic techniques to identify protein complexes. DNA chips, in conjunction with advanced genetic technologies, can be used for a systems-level understanding of transcriptional control of cell-wall synthesis and modification pathways. Epitope tagging also may be used to facilitate mRNA purification from single cells, facilitating insights into processes specific to cell types. Ultimately, GTL capabilities in systems analysis will permit an integrated systems model that can be used to support directed modification of cell walls for specific applications.

Efforts to understand and modify cell walls need to be coordinated with bioconversion and plant cell-wall deconstruction initiatives to optimize feedstock composition based on pretreatment and conversion methods and effects. These objectives also need coordination to develop analytical and visualization methods, computational facilities, and organic-chemistry methods for production of enzyme substrates and standards used in phenotyping and gene characterization.

Control of Lignin Synthesis and Structure

Although lignin is not converted readily to ethanol, lignin biomass may be amenable to chemical or thermal processing to achieve such liquid fuels as low-grade diesel or fuel oil. One aspect of optimizing biomass composition for ethanol production is minimizing lignin content. Alternatively, developing plants with modified lignin that can be removed easily during biomass processing may be possible.

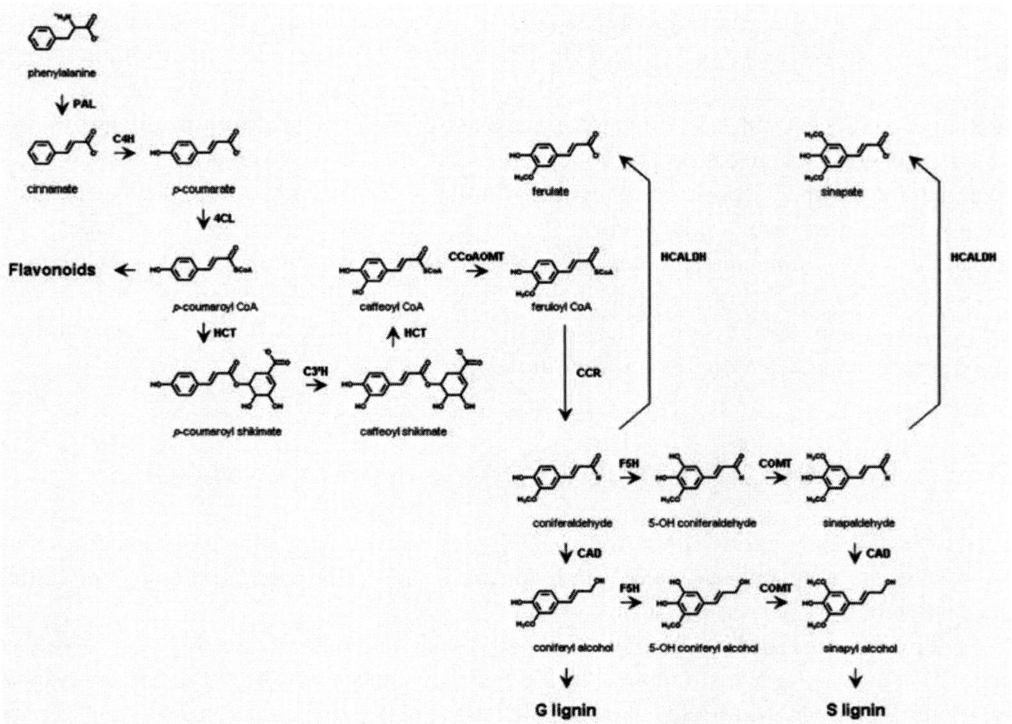

4CL, 4-(hydroxy)cinnamoyl CoA ligase; C3'H, p-coumaroyl shikimate/quinate 3'-hydroxylase; C4H, cinnamate 4-hydroxylase; CAD, cinnamyl alcohol dehydrogenase; CCoAOMT, caffeoyl CoA O-methyltransferase; CCR, cinnamoyl CoA reductase; COMT, caffeic acid/ 5-hydroxyferulic acid O-methyltransferase; F5H, ferulate 5-hydroxylase; HCALDH, hydroxycinnamaldehyde dehydrogenase; HCT, hydroxycinnamoyl-CoA shikimate/quinate hydroxycinnamoyltransferase; PAL, phenylalanine ammonia-lyase.
[Figure source: C. Fraser and C. Chapple, Purdue University]

Figure 4. Phenylpropanoid Pathway Leading to Lignin Biosynthesis in Plants.
Horizontal reactions are ring modifications; vertical reactions are side-chain modifications.

Lignin is a complex aromatic polymer associated with polysaccharides in secondary cell walls (see Figure. 3, p. 44, and Figure. 4. Phenylpropanoid Pathway Leading to Lignin Biosynthesis in Plants, this page). Lignin constitutes a significant barrier in biomass conversion to fuels by inhibiting enzyme access to polysaccharides and by releasing toxins during degradation that inhibit organism growth during fermentation of cell-wall hydrolysates to ethanol. Genetic studies have indicated that lignin reductions may cause deleterious changes in plant growth and development. However, lignin possibly may be reduced with or without harmful effects on plant growth if compensating changes could be made in the amount of cell-wall polysaccharides. Some early experiments are under way. The degree to which cellulose amount can be increased with or without simultaneous changes in hemicellulose content and composition must be ascertained.

Exploring lignin biosynthesis and its regulation in a comprehensive fashion may make possible the formulation of methods for limiting and altering lignification to maximize biomass-to-energy conversion. For instance, the gene for ferulate-5-hydroxylase has been used to increase the syringyl monomer content of poplar lignin (see sidebar, Optimizing Lignin Composition for More Efficient Bioethanol Production, p. 43). The resulting trees had normal growth and development, but the pulping time was reduced by more than 60% (Huntley et al. 2003). Similarly, opportunities exist to modify lignification cell specificity so its impact on energy conversion can be minimized. A goal is to create a lignification toolbox to manipulate polymer depositions genetically and analyze the impact of those manipulations with advanced analytical organic chemistry. Such detailed knowledge could create novel opportunities for fundamentally changing how biomass is synthesized and subsequently processed for biofuels. For instance, novel monomers might be incorporated to generate lignins with unique, useful chemistries—readily cleavable linkages that could facilitate lignin depolymerization under more benign conditions (i.e., with enzymes).

IMPROVED METHODS, TOOLS, AND TECHNOLOGIES

New analytical methods, tools, and technologies will accelerate the understanding of cell-wall synthesis, makeup, structure, and function and will speed breeding or rational modification of energy crop varieties.

At the basic research level, new and improved methods are needed to analyze wall composition and nanoscale structure. Ideally, these methods could be applied to analysis of a small number of cells. Molecules in cell walls range from 2 to 5 angstroms (0.2 to 0.5 nm) in diameter (i.e., a polysaccharide chain) and to many microns in length. Primary cell walls are from 50 to 100 nm in thickness and, in some cells, are thought to be chemically differentiated from one side to another. New imaging modalities that take advantage of various chemically specific imaging tags will support the longterm vision of in situ images of living plant cell walls. Images will reveal key molecular processes occurring in real time during the full life cycle of cell-wall formation, maturation, transformation, dehydration, and processing into simple feedstocks. The understanding obtained through research using such imaging is expected to result in quantitative, predictive modeling as a guide to formulating advanced feedstocks and their subsequent processing. A systematic approach is required to identify plant biomarkers and specific antibodies or other molecular tags useful in feedstock improvement.

Poorly understood now, the fine structure of intact walls must be studied to determine how the parts fit together to comprise the whole wall's physical properties. Some aspects of the general problem may be resolved simply by encouraging the application of such existing methods as very high resolution electron and scanning probe microscopy (see sidebar, Image Analysis of Plant Cell Surfaces at the OBP Biomass Surface Characterization Lab, p. 40).

Similarly, greater use of nuclear magnetic resonance (NMR) and magnetic resonance imaging may allow the development of 2D and 3D maps of cell-wall composition from important experimental and production species such as *Arabidopsis* and poplar. Use of NMR may be expanded by isotopic labeling and further development of solvents capable of dissolving cell-wall components. Complete annotation of 2D maps could facilitate greatly the analysis of genetic variation in cell-wall composition and the assignment of function to genes implicated in wall biosynthesis and modification.

Other approaches meriting investment include expanded collections of antibodies and aptamers to cell-wall components, the use of enzyme-based polysaccharide fingerprinting, pyrolysis gas chromatography–mass spectrometry (GC-MS), and related methods. A challenge is associated with characterizing enzymes that synthesize cell-wall polysaccharides: Many enzymes add sugars to preexisting polysaccharides (i.e., "acceptors" or "primers") that are not readily available as standards and reagents. Focused investments in carbohydrate chemistry will be required to construct substrates—including labeled substrates—for measuring the activity of many wall biosynthetic enzymes. Expertise in carbohydrate synthetic chemistry also would be a needed complement to proteomic and metabolomic capabilities envisioned in GTL capability suites. Expanded capabilities in synthetic carbohydrate chemistry could open up new high-throughput methods for characterizing carbohydrate-active enzymes based on high-density and high-diversity "glycochips." In this method, the activity or binding of a target protein could be evaluated simultaneously with hundreds or thousands of potential substrates and very small amounts of reagent.

High-throughput methods of cell-wall analysis are needed for plant breeding and improvement, allowing timely analysis with the most sophisticated analytical techniques. Ultimately, infield characterization is required to support breeding, molecular marker mapping, and studies involving such environmental variables as fertilizers and various biotic and abiotic stresses. Methods must be accurate and relatively inexpensive for the large numbers of samples typically handled during a breeding program. Additionally, they should be applicable to a wide variety of materials, from corn stover to wood. In principle, a high-throughput sample analysis may be enabled by detailed analysis of the relationship between cell-wall composition and features of Fourier transform infrared spectroscopy spectra or pyrolysis GC-MS chromatograms, combined with computational methods.

Technical Milestones

Within 5 years

- Develop rapid, accessible tools and methods for consistent biomass compositional analysis in bulk and fractions (see section, Characterizing Cell Walls Using High-Throughput Methods, p. 108).
- Identify genes for enzymes that catalyze synthesis of major polysaccharide backbones.
- Identify a substantial fraction of enzymes that catalyze synthesis of polysaccharide sidechains and determine sidechain biological function in model plant species.
- Identify enzymes that acetylate polysaccharides, and establish biological function for such modifications.
- Identify genetic regulatory factors that control lignin synthesis and deposition.

Within 10 years

- Clarify regulation of polysaccharide biosynthesis, including key steps that regulate carbon flow from photosynthesis into cell-wall polymers.
- Define mechanisms that control cellulose amount and fibril length and angle.
- Modify celluloses with altered numbers of glycan chains in secondary walls, and produce and test them in model species.
- Make available for testing biomass crop plants with decreased lignin and increased amounts of cellulose or other polysaccharides.
- Develop new tools and methods to help us understand cell-wall structure, including highly parallel computational simulations and high-sensitivity 2D NMR and MS instrumentation for analysis of lignin in small tissue samples.
- Identify all genes that catalyze synthesis of polysaccharide sidechains.

Within 15 years

- Determine regulatory genes that control amounts of major polysaccharides, including cellulose.
  Develop methods for manipulating polysaccharide composition of any particular cell type within a specific tissue.
- Make available plants with improved wall composition. These plants will have increased yields of fermentable sugars, requiring less costly preprocessing; cell-wall degradation will result in insignificant levels of inhibitory compounds (in the fermentation process).
- Develop a detailed model of lignin monomer transport, polymerization initiation, and the interactions of lignin polymers with polysaccharide components of the plant cell.

CITED REFERENCES

Huntley, S. K., et al. (2003). "Significant Increases in Pulping Efficiency in C4H–F5H–Transformed Poplars: Improved Chemical Savings and Reduced Environmental Toxins," *j Agric. Food Chem. 51(21),* 6178–83.

Pauly, M. & Scheller. H. V. (2000). "O-Acetylation of Plant Cell Wall Polysaccharides: Identification and Partial Characterization of a Rhamnogalacturonan O-Acetyl-Transferase from Potato Syspension-Cultured Cells," *Planta 210(4),* 659-67.

Understanding Biomass: Plant Cell Walls

A First Step to Optimizing Feedstocks for Fuel Production

Optimizing plant biomass for more efficient processing requires a better understanding of plant cell-wall structure and function (see next two pages). Plant cell walls contain long chains of sugars (polysaccharides) that can be converted to transportation fuels such as ethanol. The saccharification process involves using enzymes to break down (hydrolyze) the polysaccharides into their component sugars for fermentation by microbes to ethanol (see sidebar, From Biomass to Cellulosic Ethanol, p. 26). Significant challenges for efficient conversion are presented by both the large number of enzymes required to hydrolyze diverse sugar linkages and the physical inaccessibility of these compounds to enzymes because other cell-wall components are present.

Plant cell walls contain four different polymer types—cellulose microfibrils, hemicelluloses, pectins, and lignins. Microfibrils perform an important role in strengthening cell walls, thus providing support to the overall plant body. Some properties of lignin, however, interfere with enzymatic conversion of polysaccharide components. Additionally, since lignin is not readily converted to ethanol, we must find other ways it can be used if we are to maximize the yield of energy from biomass.

Several thousand genes are estimated to participate in cell-wall synthesis, deposition, and function, but very few genes have been identified and very little is known about their corresponding enzymes. Many questions remain, for example, regarding how polysaccharides and lignin are synthesized, how wall composition is regulated, and how composition relates to the biological functions of cell walls. To answer these questions, we need to discover the functions of many hundreds of enzymes, where proteins are located within cells, whether or not they are in complexes, where and when the corresponding genes are expressed, and which genes control the expression and activities of proteins involved. Application of new or improved biological, physical, analytical, and mathematical tools will facilitate a detailed mechanistic understanding of cell walls. That knowledge will permit optimization of various processes involved in producing biomass and converting it to fuels.

Major opportunities exist to increase productivity and conversion-process efficiencies by altering fundamental aspects of plant growth, development, and response to biotic and abiotic stress. Altering cell-wall composition to increase the relative amount of cellulose and to decrease lignin, for example, could have significant effects (see sidebar, Optimizing Lignin Composition for More Efficient Bioethanol Production, p. 43). Eventual development of a comprehensive physiological cell-wall model incorporating biophysical aspects with structural properties and knowledge of proteins involved will aid in rational development of highly productive feedstock species whose cell walls are optimized for conversion.

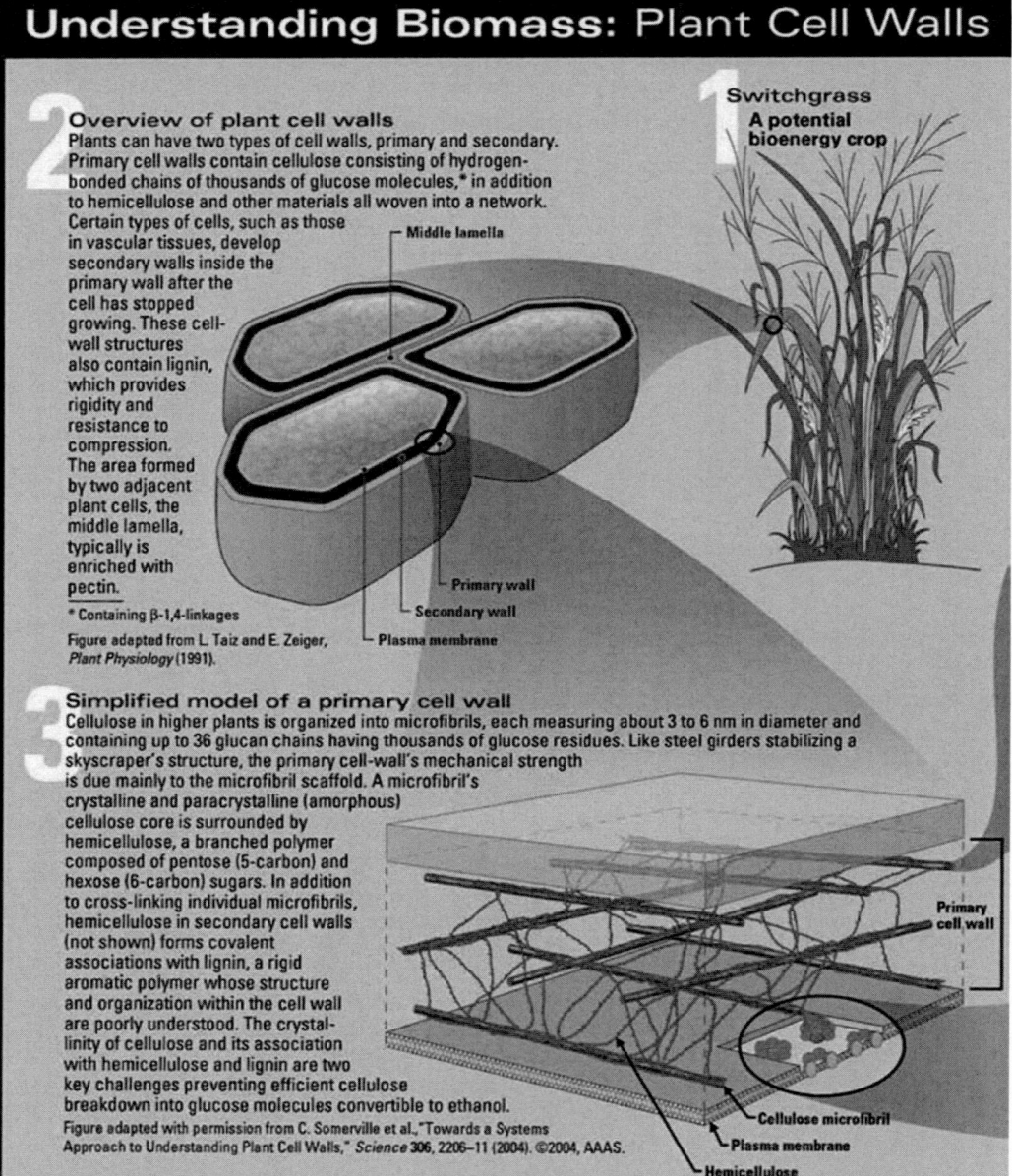

Systems Biology to Overcome Barriers to Cellulosic Ethanol

Questions Remain
- How is cellulose synthesis regulated?
- How is hemicellulose synthesized and regulated?
- How can we alter cell-wall structure (e.g., increase cellulose and hemicellulose, decrease lignin) for easier breakdown into component sugars?

7. Fragment of a cellulose molecule

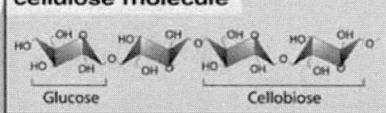

Glucose — Cellobiose

Alternating glucose residues are in an inverted orientation so the cellobiose (a disaccharide) is the repeating structural unit.

6. Crystalline cellulose

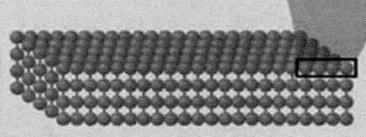

The glucan chains contain thousands of glucose residues.

5. Microfibril structure

Cellulose microfibrils are composed of linear chains of glucose molecules* that hydrogen bond to form the microfibrils.

4. Cellulose synthesis

Many enzymes involved in cell-wall synthesis or modification are thought to be located in complexes. Within the plasma membrane are rosettes composed of the enzyme cellulose synthase; these protein complexes move through the membrane during the synthesis of glucan chains (36 per rosette) that aggregate to form cellulose microfibrils. Cellulose synthase interacts with the cytoskeleton in a poorly characterized way impacting cellulose fibril orientation and perhaps length. Understanding the function of these complexes and their interactions with sugar-producing metabolic pathways will be important for eventually controlling cell-wall composition. A number of cellulose synthase genes have been cloned for a variety of plants.

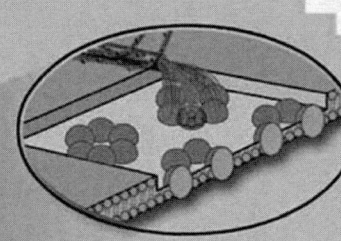

Cellulose synthase complexes

In: Biological Barriers to Cellulosic Ethanol
Editor: Ernest V. Burkheisser

ISBN: 978-1-60692-203-3
© 2010 Nova Science Publishers, Inc.

Chapter 4

FEEDSTOCKS FOR BIOFUELS

United States Department of Energy

One critical foundation for developing bioenergy crops and their processing technologies is ameliorating cell-wall recalcitrance to breakdown. Understanding cell walls is essential for optimizing their synthesis and the processes used to deconstruct them to sugars for conversion to ethanol (as discussed in the previous chapter, Lignocellulosic Biomass Characteristics, p. 39). A prerequisite for a competitive biofuel industry is the development of crops that have both desirable cell-wall traits and high biomass productivity under sustainable low-input conditions. Major agricultural crops grown today for food, feed, and fiber in the United States have not been bred for biofuels. Thus, many carefully selected traits in food and feed crops, such as a high ratio of seed to straw production (harvest index), are disadvantageous in biofuel production. A suite of new crops and new varieties of existing crops specifically bred for biofuels and adapted to a range of different soil types and climatic conditions is required.

> "Large and cost-effective energy production on a scale that significantly impacts petroleum use calls for new crops with yield and productivity not currently available...." (EERE 2003).

During the past century, improvement of agricultural crops was supported by federal investment in many aspects of basic plant science, agronomy, plant breeding, pathology, agricultural engineering, and soil science. However, many topics particularly important in biofuel production have not been emphasized and are poorly developed as a result. A recent editorial in *Science* noted:

> "There are major technological challenges in realizing these goals. Genetic improvement of energy crops such as switchgrass, poplar, and jatropha has barely begun. It will be important to increase the yield and environmental range of energy crops while reducing agricultural inputs. Plant development, chemical composition, tolerance of biotic and abiotic stresses, and nutrient requirements are important traits to be manipulated. The combination of modern breeding and transgenic techniques should result in achievements greater than those of the Green Revolution in food crops, and in far less time." (Koonin 2006)
>
> — Steven E. Koonin, Chief Scientist, BP, London

The Department of Energy (DOE) mission-oriented research program envisioned herein is designed to supplement current investment in plants, with the focus on facilitating rapid progress in formulating biomass feedstock crops, also referred to as "energy crops." One workshop goal was to identify specific areas in which a focused research investment would speed progress toward an optimized feedstock supply for conversion to biofuels. *In general terms, the goal of feedstock development is to obtain maximum usable organic carbon per acre in an environmentally and economically sustainable way.* Many previous studies have indicated that minimizing such inputs as annual field preparation and fertilization implies the use of such perennials as switchgrass and poplar, thus decreasing costs (see Figure. 1. Switchgrass Bales from a 5-Year-Old Field in Northeast South Dakota, this page). A joint document of DOE and the U.S. Department of Agriculture (USDA)—*Biomass as a Feedstock for a Bioenergy and Bioproducts Industry: The Technical Feasibility of a Billion-Ton Annual Supply* (Perlack et al. 2005)—called for perennial crops to provide about one-third of biomass- derived fuels for the initial phase of bioethanol development.

In addition, because transition to large-scale cultivation of dedicated energy crops may take years if not decades, research imperatives must be explored for optimizing the use of currently available agricultural crop and forestry residues. Sustainability will be a key issue in implementing the use of crop and forestry residues for biofuel, since the removal of crop residues can reduce organic carbon and nutrient levels in the soil and affect soil microbial community health (see section, Ensuring Sustainability and Environmental Quality, p. 68). More information about the composition and population dynamics of soil microbial communities is needed to facilitate modeling of long-term effects on soil fertility.

Figure 1. Switchgrass Bales from a 5-YearOld Field in Northeast South Dakota in 2005. Each 1200-lb. bale represents 48 gallons of ethanol at a conversion rate of 80 gallons per ton. The cultivar used in this field has a yield potential of 5 to 6 tons per acre (corresponding to 400 to 500 gallons per acre) because it was bred for use as a pasture grass. In experimental plots, 10 tons per acre have been achieved. Processing goals target 100 gallons per ton of biomass, which would increase potential ethanol yield to 1000 gallons per acre. [Source: K. Vogel, University of Nebraska]

Current knowledge indicates that perennial species expected to be used for biofuel production improve soil carbon content and make highly efficient use of mineral nutrients. Development of perennial energy crops also may facilitate use of genetically diverse mixed stands rather than monocultures of single cultivars. Because conventional crops are grown as monocultures, relatively little research has been carried out on issues associated with growing mixed stands (see sidebar, A Billion-Ton Annual Supply of Biomass, p. 10, and sidebar, The Argument for Perennial Biomass Crops, p. 59).

THE ARGUMENT FOR PERENNIAL BIOMASS CROPS

Many major agricultural crops today are annual plants propagated from seed or cuttings at the beginning of each growing season. By contrast, crops developed and grown specifically for biofuel production are expected to be based on perennial species grown from roots or rhizomes that remain in the soil after harvesting the above-ground biomass. Perennial species are considered advantageous for several reasons (see Figure. 2. Attributes of an "Ideal" Biomass Crop, p. 61). First, input costs are lower than for annuals because costs of tillage are eliminated once a perennial crop is established. Additionally, long-lived roots of perennials may establish beneficial interactions with root symbionts that facilitate acquisition of mineral nutrients, thereby decreasing the amount of fertilizer needed. Some perennials also withdraw a substantial fraction of mineral nutrients from above-ground portions of the plant at the end of the season but before harvest (see Figure. A. Nitrogen Use Efficiency Theory for Perennials, below).

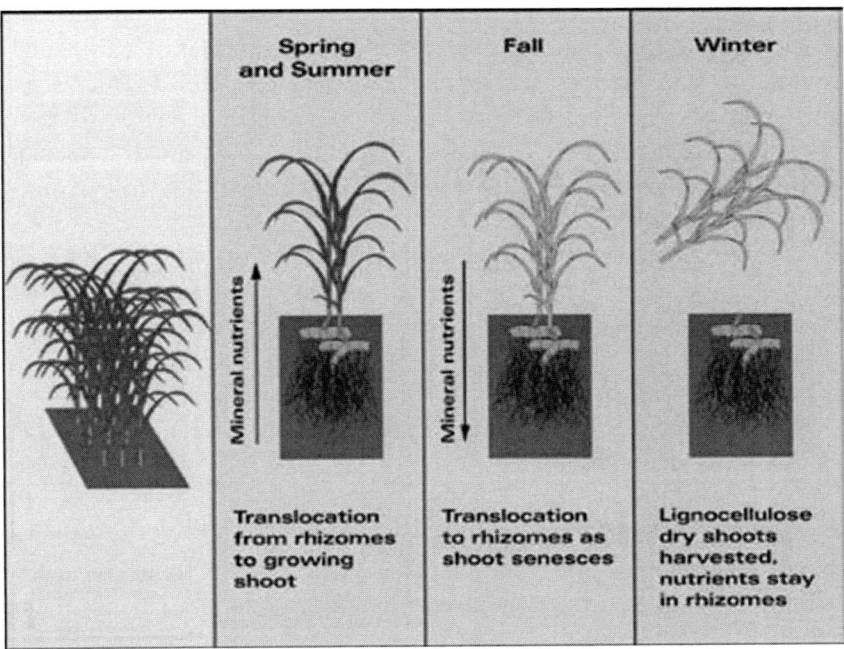

Source: S. Long, University of Illinois

Figure A. Nitrogen Use Efficiency Theory for Perennials.

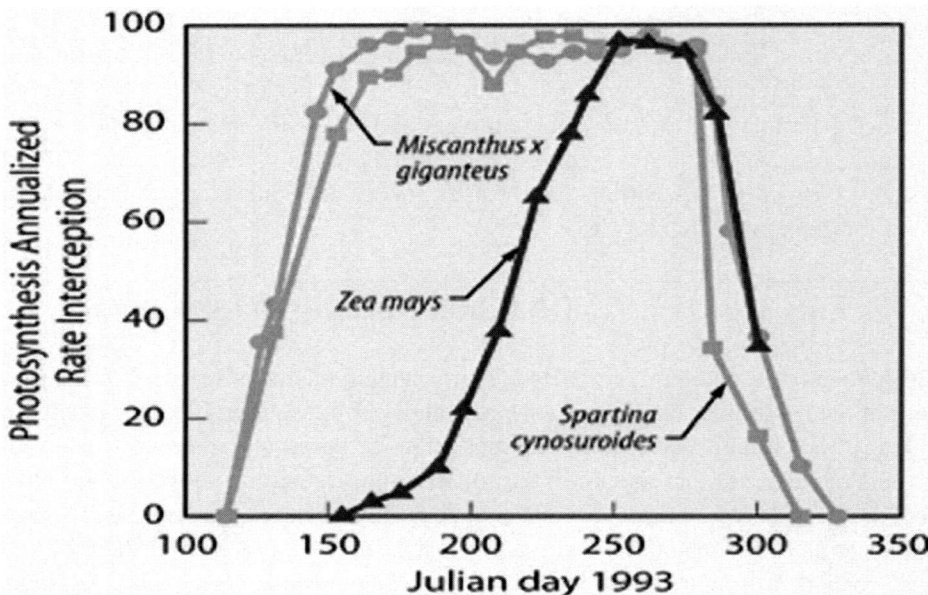

Source: S. Long, University of Illinois

Figure B. Comparing Net Photosynthesis of Corn and Several Perennial Species.
Annualized net photosynthesis is proportional to the area under the curve. Thus, if maximal rates of photosynthesis are similar, the perennial crops (yellow) have much higher annualized net photosynthesis than the annual crop, corn (*Zea mays*, blue).

Perennial plants in temperate zones also may have significantly higher total biomass yield per unit of land area than comparable annual species. Perennials establish a photosynthetically active canopy more quickly in the spring and may persist longer in the fall (see Figure. B. Comparing Net Photosynthesis of Corn and Several Perennial Species, below). Thus, their annual solar-energy conversion efficiency is higher than that of annual plants with similar capabilities.

Perennial species have much lower fertilizer runoff than do annuals. For instance, comparing the native perennial switchgrass with corn indicates that switchgrass has about one-eighth the nitrogen runoff and one-hundredth the soil erosion. Perennial grasses harvested for biomass exhibit increased soil-carbon levels and provide habitat for up to five times as many species of birds. Finally, in contrast to annual row crops that typically are monocultures, increasing habitat diversity by growing several intermixed species of perennials may prove more feasible.

Among many factors in plant productivity, several are thought to be of central importance in biofuel production. Because of evidence that most plants do not routinely achieve maximal photosynthetic CO_2 fixation rates, understanding factors limiting the overall process is important. Emphasis should be placed on determining how plants allocate recently fixed carbon to products such as storage polysaccharides (e.g., starch) and structural polysaccharides such as cellulose. Energy crops will be grown on marginal, excess, or surplus agricultural lands, so identifying factors that facilitate tolerance and survival during exposure to drought, freezing, and other abiotic stresses will be vital. Issues for perennials may be quite different from those for annuals, which have been the subject of most research and crop experimentation. These issues may be particularly acute in regard to pressure from pests and pathogens that can be controlled to some extent by crop rotation in annual species. During the past 25 years, DOE's offices of Energy Efficiency and Renewable Energy (EERE) and Basic Energy Sciences have been a primary source of research support on centrally important topics in developing bioenergy feedstocks. Research on key issues identified here constitutes a compelling opportunity. Unique capabilities developed in both the Biomass and Genomics:GTL (GTL) programs can be brought to bear on remaining practical and fundamental problems in producing feedstocks. The following sections outline these issues and opportunities in the context of tangible goals, timelines, and milestones.

CREATION OF A NEW GENERATION OF LIGNOCELLULOSIC ENERGY CROPS

Three distinct goals are associated with development of biofuel feedstocks:

- Maximizing the total amount of biomass produced per acre per year,
- Maintaining sustainability while minimizing inputs, and
- Maximizing the amount of fuel that can be produced per unit of biomass.

Exact values for each of these parameters will vary from one type of energy crop and one growing zone to another. A yield of 20 dry tons per acre per year may be considered a reasonable target in areas of the country with adequate rainfall and good soils, whereas 10 dry tons per acre per year may be acceptable in drier or colder zones.

Thus, the overall objective of developing feedstocks must be focused on broadly useful insights applicable to a variety of plant species grown under various growing conditions and exhibiting beneficial attributes (see Figure. 2. Attributes of an "Ideal" Biomass Crop, p. 61). This is best accomplished by working toward systems-level predictive models that integrate deep knowledge of underlying mechanisms for guiding cultivar and process development. This ambitious goal is only now beginning to be realized by companies that have bred advanced cultivars for major agricultural commodities.

This systems-level approach is feasible because of the last decade's biology revolution in genomic sequencing of higher plants and microbes. Sequencing provides a means for connecting knowledge about all organisms into a common framework for understanding all forms of life. Current and future DOE investments in plant DNA sequencing afford the opportunity to create the mechanistic knowledge of energy crops needed for cost-effective and practical feedstocks. This enabling information must be elaborated further by strategic investments in understanding aspects of basic biology specifically relevant to energy crops (see sidebar, Enhancing Poplar Traits for Energy Applications, p. 62).

The "Ideal" Biomass Crop?	Corn	Short-Rotation Coppice*	Perennial Grass
C4 photosynthesis	★		★
Long canopy duration		★	★
Recycles nutrients to roots			★
Clean burning			★
Low input		★	★
Sterile (noninvasive)	N/A	(★)	M.g.**
Winter standing		★	★
Easily removed	★		★
High water-use efficiency			★
No known pests or diseases			M.g.
Uses existing farm equipment	★		★

* Coppice is a grove of densely growing small trees pruned to encourage growth
** Miscanthus giganteus

Table adapted from S. Long, University of Illinois

Figure 2. Attributes of an "Ideal" Biomass Crop.

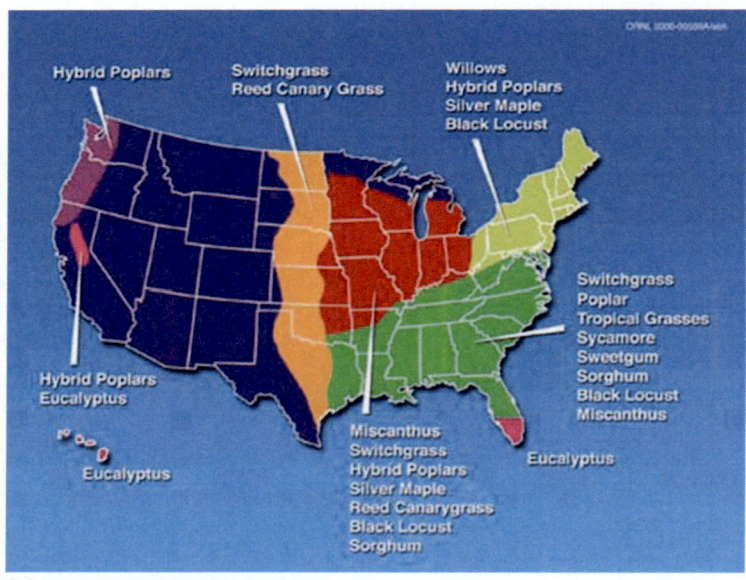

Source: Adapted from ORNL Biomass Program

Figure 3. Geographic Distribution of Biomass Crops.
Multiple types designed for various agroecosystems probably will be required to obtain enough biomass for large-scale production of liquid fuels.

Maximizing Biomass Productivity

Domestication of Energy Crops

The continental United States is composed of a number of growing zones or agroecoregions that vary with such factors as mean temperature, rainfall, and soil quality. No single plant species is optimal for all zones, so using different species as energy crops will be necessary. Previous DOE studies have identified a number of promising plant species, and academic studies have suggested additional ones (see Figure 3. Geographic Distribution of Biomass Crops, this page). In general, energy crops can be divided into two types: Those, such as maize, which are used for agricultural food and feed production but produce substantial amounts of usable biomass as a by-product (Type I), and those used only for energy (Type II). Type I plants are highly developed from many decades of research and study for another purpose (food or fiber).

ENHANCING POPLAR TRAITS FOR ENERGY APPLICATIONS

Gaining a better understanding of genes and regulatory mechanisms that control growth, carbon allocation, and other relevant traits in the poplar tree (*Populus trichocarpa*) may lead to its use as a major biomass feedstock for conversion to bioethanol. An international team led by the DOE Joint Genome Institute recently completed sequencing the poplar genome, making it the first tree (and fourth plant, after the mustard weed *Arabidopsis thaliana,* rice, and the alga *Chlamydomonas*) to have its complete genome sequenced (Tuskan et al., in press). These data now offer the molecular access needed to define, quantify, and understand—at a mechanistic level—basic biological processes that impact important traits.

Poplar was chosen for sequencing because of its relatively compact genome (500 million bases), only 2% that of pine. Moreover, many species are available worldwide, and their rapid growth allows meaningful measures of important traits within a few years. Extensive genetic maps already available include initial identification of markers associated with such traits.

Early comparative sequence analyses of poplar and *Arabidopsis* genomes are providing insights into genome structure and gene-family evolution; biosynthetic processes such as cell-wall formation, disease resistance, and adaptation to stress; and secondary metabolic pathways. Comparisons of gene-family sizes show substantial expansion of poplar genes involved in carbon to cellulose and lignin biosynthesis.

Moving from a descriptive to predictive understanding of poplar growth, development, and complex function will require integration of sequence information with functional data. These data will be generated by such new tools and approaches as gene and proteome expression studies, metabolic profiling, high-throughput phenotyping and compositional analysis, and modeling and simulation. Ultimately, this information will lead to the engineering of faster-growing trees that produce more readily convertible biomass (see below, Figure A. Vision for the Future). The International Populus Genome Consortium has produced a science plan to guide postsequencing activities (*The* Populus

Genome Science Plan 2004–2009: From Draft Sequence to a Catalogue of All Genes Through the Advancement of Genomics Tools, www.ornl.gov/ipgc).

Other areas to be addressed for poplar and other potential bioenergy crops include sustainability for harvesting biomass, harvesting technologies to remove biomass at low costs, and infrastructure technologies that allow biomass to be transported from harvest locations to conversion facilities.

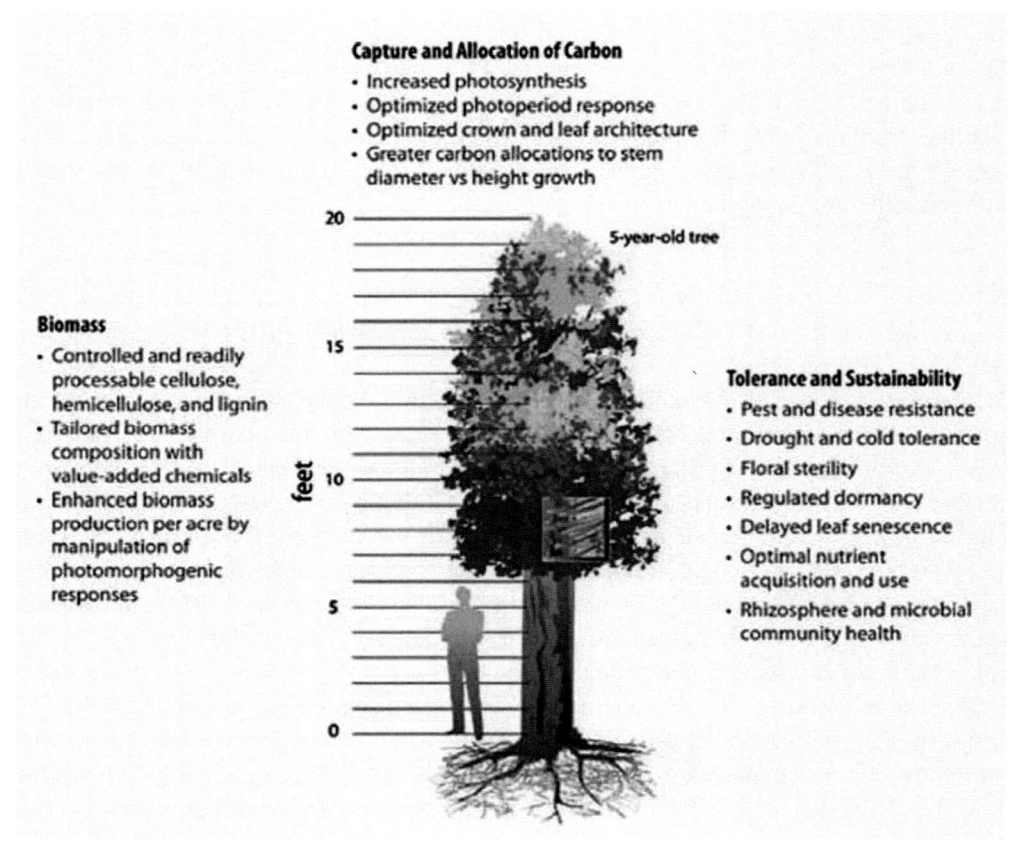

Figure A. Vision for the Future: Desired Traits of the Domesticated Energy Poplar.

They were bred as monocultures under intensive agriculture. As noted elsewhere in this document, research priorities regarding production of energy from these crops are concerned largely with three issues. These issues are sustainability following removal of previously underutilized straw or stover, processing the residues, and the possibility of making improvements in cell-wall composition for enhanced conversion to fuels.

By contrast, Type II plants are relatively poorly developed as agronomic crops and generally are perennials. Perennial herbaceous and woody plants have several properties that make them better suited for biofuel production than are annual crops (see sidebar, The Argument for Perennial Biomass Crops, p. 59). First, because they typically retain a significant tissue mass below ground, they rapidly form a canopy in the spring and accumulate biomass when many annuals are still seedlings. Thus, they may exhibit higher rates of net photosynthetic CO_2 fixation into sugars when measured annually, resulting in higher amounts of total biomass accumulation per acre per year. Second, perennials require little if any tillage, saving energy and labor and significantly reducing soil erosion and nutrient loss. Perennials such as switchgrass and *Miscanthus* can be harvested annually without replanting. Third, perennials typically withdraw mineral nutrients into roots at the end of a growing season, thereby reducing fertilizer costs. Perennial herbaceous and woody plants represent a critical component of our bioenergy future.

Advances in conventional breeding, coupled with molecular tools and high-throughput transformation systems (Busov et al. 2005) will be required to accelerate domestication of species having promise as energy crops. Associating genotypes with phenotypes will require high-throughput technologies for genotyping (i.e., identifying genes and alleles) and phenotyping a wide array of relevant traits (e.g., biomass yield, cell-wall composition) with "molecular" techniques such as gene expression (transcriptomics), proteomics, and metabolomics. Availability of whole-genome sequences and deep expressed sequence tag (EST) libraries of all potential energy crops will be necessary to provide genetic reagents for marker-aided selection, association genetics, and transformation studies (see sidebar, Marker-Assisted Breeding, p. 64). Association genetics identifies allelic variants [e.g., single nucleotide polymorphisms (SNPs)] with empirically determined phenotypic effects on traits of interest and allows selection for favorable alleles using molecular markers (Rafalski 2002; Neale and Savolainen 2004; Remington et al. 2001). This form of marker-assisted selection is direct (as opposed to indirect on quantitative trait loci) and allows simultaneous selection of many genes. The result is a rational, marker-aided breeding and selection approach with the expectation of significantly enhanced genetic gain and an accelerated development process (Brown et al. 2003).

For most Type II biomass crops, significantly enhanced breeding, testing, and selection populations will be necessary, along with appropriate infrastructure to breed for desired traits and adaptability across a wide array of environments in multiple physiographic regions. In addition to targeted breeding, many crops will require fertility control in the field, either to ensure parentage or prevent gene flow to wild populations. A holistic approach for improving feedstocks includes molecular resources, high-throughput screening tools, and well-characterized breeding populations. GTL resources and technologies are well suited to creating comprehensive sets of molecular markers (i.e., SNPs and single sequence repeats) and high-throughput, low-cost phenotyping tools. These efficient deliverables will be used to develop appropriate cultivars and varieties to meet target goals. GTL genomic and other resources will play critically important roles via implementation of deep EST sequencing, marker identification, high-throughput genotyping, and development and application of analytical tools .These tools will be used for high-throughput molecular phenotyping of biomass composition and plant structure and a high-throughput transformation process for major biomass energy species. Such capability suites also will provide centralized bioinformatics support for analysis and archiving of genome data.

By definition cultivar development, field and plantation establishment and growth, and biomass conversion to biofuels necessitate a holistic, systems biology approach. Integrating the smaller subcomponents will be challenging, requiring a coordinated and focused program to facilitate exchange of information and genetic materials across organizations and institutions.

> ## MARKER-ASSISTED BREEDING
>
> One of the ways in which DNA sequencing of biomass species impacts biofuel feedstock development is by enabling accelerated breeding methods to be applied to plants used for biomass production. The goal in plant breeding is to first identify useful genetic variation for traits of interest (e.g., disease resistance and drought tolerance) by screening natural or mutagenized populations of individuals. Many such traits are controlled by multiple genes. Individuals with useful variations are intercrossed to produce progeny with new combinations of the useful variation. Usually, many traits are of interest and many sources of variation are used, making the overall breeding process very time consuming and expensive. The availability of large amounts of sequence information facilitates identification of DNA polymorphisms—small differences in the DNA sequences of individuals within a species. Having complete genomic sequences also makes possible the identification of genes located near polymorphisms on chromosomes.
>
> This knowledge has practical applications in "marker-assisted breeding," a method in which a DNA polymorphism, closely linked to a gene encoding a trait of interest, is used to track the trait among progeny of sexual crosses between plant lines. The method allows breeders to monitor plants for a trait that may be expressed only in certain tissues or developmental stages or may be obscured by environmental variation. Similarly, by correlating traits and DNA polymorphisms in individuals from genetically diverse natural populations, associating a particular chromosomal region with a trait of interest frequently is possible. If a large number of polymorphisms are available, the amount of time required to breed an improved plant cultivar is greatly reduced.
>
> In principle, plants with optimal combinations of parental genes may be identifiable within the first several generations following a sexual cross rather than eight generations or more following conventional breeding. In the case of species such as trees, marker-assisted breeding could eliminate many decades of expensive steps to develop more highly productive plants.

Enhancing the Yield of Biomass Crops

The yield of biomass crops can be defined as the amount of fixed carbon per acre per year. Achieving the maximal yield of a dedicated energy crop (Type II) is a significantly different goal from maximizing the yield of most existing crop species (Type I), where only the number of reproductive or storage organs is considered. The yield of a Type II species is a function of the total number of cells per acre multiplied by the mean amount of accumulated carbon per cell. Thus, biomass yield can be enhanced by increasing the number of cells per acre per year, the amount of carbon per cell, or both. Achieving either type of enhancement is a complex systems problem. At the core of the problem, however, is the need to maximize photosynthetic CO_2 fixation to support carbon accumulation. Additionally, fixed carbon must be directed into either cell-wall polymers or storage carbohydrates or used to support extra cell division. Cell-wall polymers include cellulose, hemicellulose, and lignin; storage carbohydrates include sugars and starches.

Plants are regulated to fix only the carbon needed for normal growth and development. This generally is referred to as "source-sink" regulation, a poorly understood phenomenon. Plants can fix considerably more carbon, however; the actual photosynthetic CO_2 fixation rate of most or all plants is significantly below (i.e., ~50%) the rate observed following experimental partial defoliation. That is, plants appear to accumulate more carbon per unit of leaf area following defoliation than they normally would without any changes in architecture or photosynthetic electron transport. When mechanisms underlying this regulation are understood, plants can be developed that exhibit significantly higher rates of net photosynthetic CO_2 fixation and higher amounts of total carbon accumulation per acre per year. Therefore, a high-priority research goal is to understand mechanisms that regulate net photosynthetic CO_2 fixation. A closely related priority is to identify factors that limit carbon flux into cell-wall polysaccharides and storage polymers.

A complementary approach is to identify factors that regulate plant growth rate and duration. Different plant species vary widely in growth rates, suggesting that growth rates are under genetic control and, therefore, subject to modification. Recently, several genes have been identified in functional genomics screens that cause significant increases in growth rates of different types of plants. Identifying other genes that control growth and development and understanding gene action may create new opportunities to develop highly productive energy crops (see Figure. 4. Growth Rate Modification, this page).

Enhancing Abiotic Stress Tolerance of Biomass Species

Water availability is a major limitation to plant productivity worldwide, generally in two ways. First, because water escapes from plant leaves through stomata when CO_2 enters, a certain amount of water is required to support a unit of photosynthetic CO_2 fixation. Plants with C4 photosynthesis (e.g., corn, sugarcane, switchgrass, and Miscanthus) typically require less water per unit of CO_2 fixed than do C3 species (e.g., wheat and soybean) because C4 plants can achieve high rates of CO_2 fixation with partially closed stomata. Other plants such as cacti close their stomata during the day to reduce water loss but open them at night to take in CO_2 for photosynthesis the next day. This phenomenon also inhibits carbon loss by photorespiration. Except for the possibility of enhancing these adaptations, current theory implies no other options for significantly reducing a plant's water requirement to obtain maximal yields.

The water problem's second component, however, concerns the effects of temporal variation in soil-water content. In rain-fed agriculture, periods of low soil-water content are frequent because of irregularities in rainfall. The ability of plants to survive extended periods of low soil water can be a critical factor in a crop's overall yield (see Figure. 5. Corn Yield on a Missouri Experiment Station, this page). Furthermore, different plants exhibit widely different abilities to survive extended periods of drought, indicating that drought-tolerant energy crops may be possible (see Figure. 6. Modification in Drought-Stress Tolerance, this page).

Source: Mendel Biotechnology

Figure 4. Growth Rate Modification.
The *Arabidopsis* plant on the right has been modified by altering the expression of regulatory genes controlling growth.

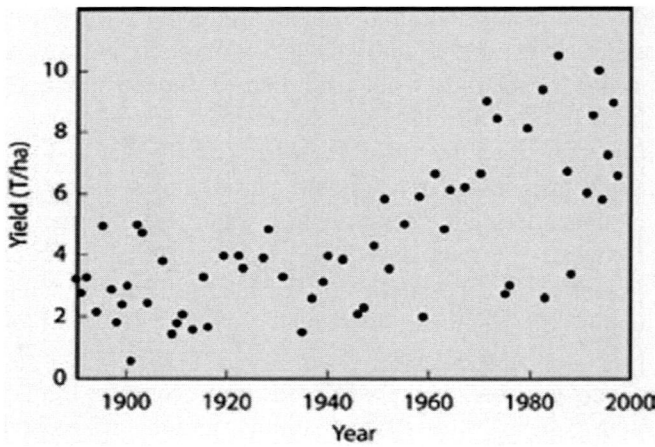

Figure 5. Corn Yield on a Missouri Experiment Station.
This figure illustrates wide differences in yield observed in various annual growing conditions, including periods of drought. The strong upward trend in yield reflects advances in breeding and agronomical practices for corn. [Figure adapted from Q. Hu and G. Buyanovsky, "Climate Effects on Corn Yield in Missouri," *j Appl. Meteorol.* 42(11), 1623–35 (2003).]

Currently, the most productive farmland is used for food production, with an amount held in reserve [Conservation Reserve Enhancement Program (CREP) lands]. While initial energy crops would be grown in highly productive CREP land, a substantial proportion of biomass crops will be grown on marginal land that is suboptimal in water availability, soil quality, or both. Recent progress in understanding the mechanistic bases of plant drought, salt, and cold tolerance has raised the possibility of modifying plants to enhance productivity under these and other stress conditions. A priority in dedicated energy crops is to understand mechanisms by which plants survive drought and adapt this knowledge to improving energy crops.

Figure 6. Modification in Drought-Stress Tolerance.
The plants on the left of each photograph are unmodified, and the plants on the right have been altered genetically for drought-stress tolerance. All plants were subjected to a severe drought. [Photos ©2003 Monsanto Company. Used with permission.]

Understanding and Eliminating Undesirable Biomass Crop Characteristics

Research should be conducted to understand and eliminate such undesirable plant traits as inappropriate residues and invasiveness of non-native energy-crop species. Two examples are described here.

Relatively large amounts of silica in plants such as rice lead to accumulations of ash when the plant biomass is burned for energy. When the biomass is converted to liquid fuels, unusable solids are produced. Since many plants do not accumulate silica, the trait appears to be dispensable; developing cultivars with reduced silica accumulation by genetic methods may be possible. The roles of silica in plant growth and development, however, are poorly understood and need further investigation.

Many features considered ideal for a biomass crop (see Figure. 2, p. 61) are characteristic of invasive weeds, particularly perennial C4 grasses. Thus, a key consideration in adapting these grasses for use as dedicated energy crops is to ensure that the species can be contained and will not become a problem. Some highly productive perennial grasses, such as *Miscanthus giganteus,* have been studied intensively in Europe for more than a decade and are thought not to exhibit invasive characteristics. All candidate energy crops, however, should be studied directly for potential invasiveness at diverse locations within the United States. These studies also can provide insights into pests and pathogens that might pose a threat to productivity.

Technical Milestones

Within 5 years

- Facilitate the initiation and advancement of biomass breeding programs in key U.S. geographic regions in cooperation with USDA, private companies, and universities.
- Develop appropriate test populations for conducting association genetics and quantitative trait locus identification (QTL, direct and indirect marker-assisted selection).
- Identify and target for selection and improvement key traits that affect biomass yield and conversion efficiency.
- In support of marker development for gene discovery, carry out targeted sequencing (SNPs, SSRs, ESTs, and cDNAs) for potential biomass species having large and complex genomes; sequence whole genomes for species with modest genome sizes.
- Develop markers that can differentiate superior parents and offspring.
- Initiate and validate methods for high-throughput screening for specific traits such as improved cell-wall digestibility.
- Formulate the infrastructure for functional screening of thousands of potentially useful genes in selected species.

Within 10 years

- Apply high-throughput phenotyping tools to integrated conventional and molecular breeding programs.
- Gain new understanding of genome structure and gene expression in bioenergy crops, including the genetic basis of heterosis (hybrid vigor). Major barriers to facile transformation of select genotypes will be overcome.

Within 15 years

- Identify and integrate major new domestication genes into energy crops.
- Integrate enabling technologies with breeding programs to allow for deployment of genetically superior materials over large land bases dedicated to biomass production. Time to commercial deployment will be shortened dramatically through highly reliable screening methods, genetic control of flowering and sexual reproduction, and effective early-selection models for predicting performance and yield. Genetically enhanced cultivars, hybrids, and varieties capable of meeting targeted goals for ethanol yield per acre per year will be available for major biomass crop species.

ENSURING SUSTAINABILITY AND ENVIRONMENTAL QUALITY

To ensure the viability of bioethanol to meet the large national need for transportation fuels, we must understand the effects of long-term biomass harvesting on soil fertility and other aspects of sustainability. Objectives for this work are to determine how to maintain soil ecosystem function and productivity. A further objective is to develop management practices that can optimize sustainability. Because very long periods of time may be required for directly testing the effects of biomass removal on soil quality, a mechanistic understanding of this issue is essential to obtaining predictive models and monitoring procedures. The critical question in using crop residues and dedicated crops for biomass energy is, How much, if any, of above-ground biomass needs to be left on a field to protect soil from erosion and sustain soil function (soil quality)?

In the past 30 years, crop and soil research has emphasized using crop residues in minimum and no-till farming operations to improve soil organic carbon (SOC) and soil quality. Research has demonstrated that corn stover removal in some production systems can reduce grain yield (Wilhelm et al. 2004). The amount of stover removal and lower yield were associated with the amount of SOC (Maskina et al. 1993). At any time, SOC content is the balance between the rates of input and decomposition (Albrecht 1938). If all other cultural practices are unchanged, removal of crop residue will further decrease carbon inputs and SOC will decline (Follett 2001) (see Figure. 7. Soil Carbon Alterations with Management Changes, p. 69). Loss of SOC typically has detrimental effects on soil productivity and quality, presumably because microbial communities are impacted negatively. For example, considerable carbon is translocated to mycorrhizal fungus communities that play a significant role in soil nutrient cycling (Fitter et al. 2005).

This research must be completed for each major agroecosystem in the United States where crop residues and dedicated crops are a feasible supply source for biomass energy. Sustainability analysis of both dedicated energy crops and of stover or straw removal from agricultural crops should be carried out since the effects may be significantly different.

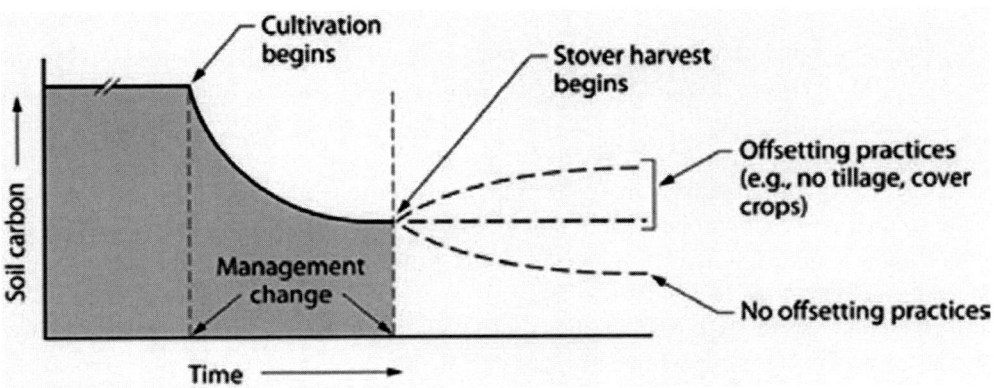

Source: W. W. Wilhelm, USDA Agricultural Research Service, Lincoln, Nebraska.

Figure 7. Soil Carbon Alterations with Management Changes.
Cultivation generally leads to reduction of soil organic carbon, which, without offsetting practices, is exacerbated by corn-stover removal. A mechanistic understanding of long-term harvesting effects on soil fertility and other aspects of sustain- ability is essential for predictive models and monitoring procedures.

Soils are complex ecosystems composed of unknown numbers of organisms. The effects of crop-residue quantity and composition on soil ecosystems are largely unknown. Soil microbiologists need access to large-scale genomics and other analytical facilities to conduct soil microbial research. Systems biology analyses of highly productive soils, including their ecosystem-scale genomic characterization, are accessible with current sequencing capabilities. Questions include the following.

- What characteristics and functions of soil microbial communities are needed to maintain soil-ecosystem function and productivity?
- How do soil microbial communities function?
- What are the interactions, positive and negative, among microbes, fungi, and roots in the rhizosphere?
- How do nutrient levels affect microbial communities, and how do microbial communities affect nutrient availability?
- How can management of microbial communities improve productivity?
- How much carbon from crop residues and dedicated energy crops is needed to maintain soil ecosystem function and productivity?
  Does the composition of residue have an effect?
  What is happening both physically and biologically in soils as various levels of plant biomass are removed?
- Can management factors, no tillage, modified minimum tillage, or special practices affect the amount of needed residue?
- How does soil composition vary between annual and perennial crops in response to varying levels of biomass removal?
- What are the microbial-community characteristics of marginal and severely depleted lands, and how can they be restored to support energy crops?
- Can energy crops be used to restore lands for food and fiber crops by building up carbon and nutrients in marginal lands? Finally, how might these processes be utilized for reduction in greenhouse gas emissions or carbon sequestration?

GTL facility capabilities can be used in many ways to answer these general questions. In particular, GTL facilities could support ecosystem-analysis studies, including sequencing of microbial communities and genomics approaches for analysis of microbial-community functioning. GTL could support technologies to evaluate how microbial communities function by, for example, developing gene chips with millions of diagnostic probes.

Determination of microbial-community physiology must be linked with analysis of soil physicochemical states—facilitating analysis of soil emissions or carbon sequestration in soils. Improved methods of soil carbon analyses need to be developed and made available in high-throughput format. Current methods are extremely laborious. Improving analytical methods for analyzing composition of plant cell walls, described below, also may facilitate analysis of soil carbon composition.

Technical Milestones

Within 5 years

- Determine the effects of corn-stover and other crop-residue removal on soil productivity in each major agroecosystem of the United States.
- Use existing study sites where stover has been removed for 5 or more years to initiate comprehensive analysis of microbial communities (and soil carbon and mineral content) in underlying soils, in comparison with land where stover has not been removed.
- Establish long-term study sites for future sustainability studies on prospective energy crops in lands in all major agroecosystems.
- Pursue long-term contracts with private or public providers. The European *Miscanthus* Productivity Network, supported by the European Commission, provides a useful model. Sequence analysis of DNA extracted from soils should be carried out to survey diversity (Handelsman et al. 1998). Sequence analysis should be used to develop diagnostic methods for examining the dynamics of soil microbial composition and abundance.

Within 10 years

- Conduct additional studies with varying harvest rates and management practices, including cropping systems and analyses of microbial communities and of carbon levels in the soils.
- Perform comprehensive ecological system analyses of soils as a living medium.
- Include not only microorganisms present but also how their complex interactions contribute to their net function as a community. Markers of optimally functioning microbial communities should be developed to enable identification of sustainability requirements for soils in each major agroecosystem. In addition to its value to the energy-production function of croplands, this research should enhance the sustainability of food production.

Within 15 years

- Formulate management guidelines on the amount of biomass that can be removed in each major U.S. agroecosystem.

MODEL SYSTEMS FOR ENERGY CROPS

Application of model systems toward the study of both basic and applied problems in plant biology has become routine and can quickly bring the techniques of 21st Century systems biology to bear on the complex problems associated with domesticating energy crops (see sidebar, Translational Research: The Path from Discovery to Applications, p. 72). Researchers using the model dicot *Arabidopsis thaliana* have made tremendous strides in understanding areas of plant biology ranging from nutrient uptake and metabolism to plant-pathogen interactions. Unfortunately, as an annual dicot, *Arabidopsis* is not an optimal model to study questions unique to potential woody and grassy perennial energy crops (e.g., wood formation in trees or cell-wall composition in grasses). Despite its sequenced genome and genetic resources, rice also is not an ideal model for grassy perennial energy crops because it is a specialized semiaquatic tropical grass. In addition, its large size, long generation time, and demanding growth requirements make experiments expensive. Using *Brachypodium* and *Populus* as model systems would provide researchers working to domesticate energy crops with some of the most powerful tools developed by the highly successful *Arabidopsis* community. These model systems would help identify genes controlling traits relevant to energy-crop productivity and quality, including such global processes as cell-wall biosynthesis, nutrient uptake, carbon flux, and plant architecture. The models can be used for rapid testing of strategies to improve the usefulness of grasses and trees as energy crops. Such tools would allow scientists to use both forward and reverse genetic approaches and modern molecular genetic methods to identify genes controlling traits relevant to the design of superior energy crops. This is important because of the problems (i.e., difficult transformation, large size, long generation time, and self-incompatibility) associated with working directly with energy crops.

Brachypodium distachyon is a small temperate grass with all attributes needed to be a modern model organism, including simple growth requirements, fast generation time, small stature, small genome size, and self-fertility (Draper et al. 2001). *Brachypodium* also is transformed readily by *Agrobacterium* (Vogel et al. 2006) or biolistics (Christiansen et al. 2005), thus facilitating many biotechnological applications. *Brachypodium* is now scheduled to be sequenced by the DOE Joint Genome Institute. The *Populus* genome has been sequenced (Brunner, Busov, and Strauss 2004; Tuskan, DiFazio, and Teichmann 2004), assembled, and annotated through investments from DOE, NSF, and a large international consortium (Tuskan et al., in press). This resource is available to researchers working on a dedicated woody crop for biofuels and biomass energy applications. Despite these advances, many molecular tools and resources currently not available would greatly enhance the discovery and use of genes and gene families related to energy traits and the development of poplar as an energy crop (see sidebar, Enhancing Poplar Traits for Energy Applications, p. 62). As with other model organisms, research on poplar would be facilitated by organism-specific tools such as full-genome DNA chips, sequence-indexed mutations, facile transformation methods, and specialized libraries for methods including two-hybrid screens.

TRANSLATIONAL RESEARCH: THE PATH FROM DISCOVERY TO APPLICATIONS

Model organisms play an important role in both basic and applied research. The most highly developed plant model is *Arabidopsis thaliana*, a small plant in the mustard family (Fig. 4. Growth Rate Modification, p. 65). It was adopted as a model in the early 1980s because of its rapid life cycle, simple genetics, small genome (125 Mb), easy transformability, and traits typical of flowering plants in most respects. Genome sequencing was completed in 2000 by an international consortium, and about 13,000 researchers worldwide currently use the plant for studies of plant biology.

The *Arabidopsis* community has developed a suite of very powerful experimental resources that include several hundred thousand sequenced insertion mutations available through stock centers, a full-genome DNA chip that can be used to measure the expression of most genes, a high-density polymorphism map, and a heavily curated database (arabidopsis.org) that provides access to genomic information. These and other resources have resulted in an explosion of fundamental discoveries about all aspects of plant growth and development. In addition, several companies, founded to exploit knowledge gleaned from *Arabidopsis*, have translated the new information into such crop species as corn, soybean, and canola (see Figure A. Drought Resistance in Canola Caused by Directed Modification of a Single Gene, below). Plant-improvement goals that were unattainable by applied research on crop plants have been realized by translating basic research on model species to crops.

Much accumulated knowledge about crop species and model organisms will be applicable to improving species suited for use as dedicated biomass crops. However, some traits of interest in that regard have not been a priority in crop research and are poorly understood. For instance, knowledge about cell-wall structure, function, and synthesis is very underdeveloped and will need extensive research. One important species suited for use as a biofuel is poplar, a model woody species that has a relatively small genome. Poplar was sequenced recently at the DOE Joint Genome Institute (see sidebar, Enhancing Poplar Traits for Energy Applications, p. 62).

The DNA sequence of many poplar genes is similar to that of corresponding *Arabidopsis* genes. This comparability will help link mechanistic knowledge about these species to a holistic understanding of common processes. Poplar is relatively easy to transform, which greatly facilitates experimental tests of theories about protein and gene function. Most important, poplar has several important traits, such as a long life cycle and the formation of wood, that cannot be studied in herbaceous plants.

Figure A.. Drought Resistance in Canola Caused by Directed Modification of a Single Gene. [Photo ©2003 Monsanto Company. Used with Permission.]

> *Brachypodium distachyon* is another potentially important model for such highly productive grasses as switchgrass and *Miscanthus*. Interest in *Brachypodium* arises because its very small genome has a DNA content about 2.5 times larger than that of *Arabidopsis*. Additionally, its simple growth requirements, small stature, self-fertility, and ready transformability make it well suited to become a model organism. Because *Arabidopsis* is distantly related to and differs from grasses in a number of important respects (e.g., cell-wall composition), *Brachypodium* could become a powerful new model for cell and molecular biological studies of grasses. A high priority in facilitating the development of *Brachypodium* is sequencing its genome.

Brachypodium and *Populus* must be accelerated into powerful model systems by first sequencing *Brachypodium* and then creating extensive EST databases (i.e., 1 to 2 million per species) from diverse (including full-length) cDNA libraries for both organisms (Sterky et al. 2004). *Agrobacterium*-mediated transformation of both *Brachypodium* and *Populus* has been developed but must be optimized for high-throughput approaches to gene discovery. High-throughput capacity for transforming diverse genotypes would enable forward and reverse genetic approaches by generating collections of sequence-indexed insertional mutants, including but not limited to activation tagging (Groover et al. 2004, Busov et al. 2005), and by allowing targeted silencing or overexpression of large numbers of genes. Transposon-based methods for generating insertional mutants also should be developed. High-throughput resequencing capacities should be used for identifying molecular markers (e.g., SNPs) associated with biomass and bioenergy traits. Similarly, phenotyping (transcript and protein profiling) capacities that identify genotypes associated with relevant cell-wall composition or agronomic traits would be useful.

Using this suite of tools, the function of all *Brachypodium* and *Populus* genes will be evaluated for their contribution to biomass energy-relevant traits. Genes deemed relevant will be used to create advanced genotypes in energy crops. Public organism-specific databases should be created and curated in parallel with advanced experimental capabilities for these organisms.

To set research priorities, the DOE Office of Science's GTL program is interacting with the conversion program managed through EERE's Office of the Biomass Program and other interested parties (e.g., USDA, NSF, and private industry).

Technical Milestones

Within 5 years

- Complete *Brachypodium* genome sequencing, plus EST sequencing and molecular-marker identification for both *Brachypodium* and *Populus*.
- Optimize transformation methods for diverse genotypes of *Brachypodium* and *Populus,* and generate sequence-indexed insertion mutant populations of >100,000 events for both organisms.
- Develop tools for transcriptional and protein profiling and carry out first-stage profiling of key tissues (e.g., cambial development in *Populus*).

- Develop organism-specific databases around annotated genome sequences.

Within 10 years

- Identify genes relevant to biomass production (e.g., those that control cell-wall composition, nutrient uptake, carbon partitioning, flowering, stress tolerance, and disease resistance) using forward and reverse genetic screens and natural populations based on high-throughput phenotyping [e.g., use Fourier transform infrared spectroscopy (called FTIR) or other techniques that can be automated to identify genotypes with altered cell-wall composition and advanced image analysis to examine plant architecture].
- Develop and test strategies using these genes to improve energy crops in *Brachypodium* and *Populus*.

Within 15 years

- In breeding programs, use genes and strategies shown to affect biomass production and quality in *Brachypodium* and *Populus* to improve energy crops, either by transgenic approaches or by marker-assisted selection of naturally occurring variability in orthologous genes.
- Transfer improved energy crops and genetic information to the conversion program managed through EERE and the Office of the Biomass Program and to other interested parties (e.g., USDA, NSF, and private industry).

THE ROLE OF GTL CAPABILITIES FOR SYSTEMS BIOLOGY

Capabilities of GTL (within DOE's Office of Biological and Environmental Research) will support or enable the attainment of lignocellulosic characteristics (see Lignocellulosic Biomass Characteristics chapter, p. 39) and feedstock objectives in numerous ways. Potential contributions of each of these, whether distributed or consolidated in a facility, are described below.

Protein Production Capabilities

Most enzymes of interest in cell-wall biosynthesis are thought to be membrane associated and, therefore, difficult to purify and characterize by conventional methods. Based on preliminary experiments with membrane-localized glycosyltransferases, a potentially powerful tool in characterizing the function of all plant glycosyltransferases is to express them in suitable hosts such as insect cells, in which endogenous activities will not interfere with assays. This capability could undertake the high-throughput expression of all glycosyltransferases and other cell wall–active enzymes (e.g., peroxidases and laccases) from higher plants (e.g., *Arabidopsis, Populus, Brachypodium*, or rice) or specific biomass species in one or more suitable surrogate hosts. Such expression systems will greatly facilitate function identification of enzymes that catalyze synthesis of cell walls. Heterologously expressed proteins would be tested for their substrate specificity and catalytic activity and also will be used as a resource for generating antibodies and tags.

Antibody resources would have multiple applications. A collection of fluorescent- or epitope-tagged proteins in a plant such as *Arabidopsis* or *Brachypodium* also would be very useful for a variety of reasons. First, the tagged proteins could be used to recover complexes for assay by mass spectrometric and imaging methods. Fluorescent tags could be utilized for cellular localization or collocation [e.g., confocal microscopy or fluorescence resonance energy transfer (called FRET)]. Additionally, if certain proteins are not active in a surrogate, the tagged version in a plant host might be used to identify function by in vitro assays of affinity-purified proteins. Developing such a resource would require a high-throughput transformation capability that also would be useful for generating insertion mutants in *Brachypodium* and *Populus*.

Molecular Machines Capabilities

Cellulose is synthesized by an intricate multienzyme complex in the plasma membrane, and matrix polysaccharides probably are synthesized by multienzyme complexes in the Golgi. Understanding the function of these complexes and their interactions with metabolic pathways that produce sugar nucleotides in the cytosol will be important for understanding the control of polysaccharide biosynthesis. In addition, cellulose synthase interacts with the cytoskeleton to control the orientation of cellulose fibrils and possibly influence fibril length. Many questions about this and related processes will benefit from the application of new technologies for imaging and manipulating protein complexes. Since most or all such complexes are membrane bound, specialized methods must be developed to work with the complexes. Also, because plant cell walls are intricate mechanical assemblies, tools developed for investigating protein machines may be used to gain new insights into assembly and function of the large polysaccharide complexes that comprise cell walls.

Proteomic Capabilities

These capabilities will facilitate identification of proteins that make up complexes in such model plants as *Arabidopsis, Populus,* and other biomass crops. Since many complexes are membrane bound, innovative methods will be needed to purify membrane complexes using an array of capabilities. Also, because of poor correlation between mRNA and protein abundance, proteomic capabilities will enable documentation of the proteome of living cell types found in vascular and other tissues, including ray parenchyma and phloem, as a function of time and conditions. Proteomic analysis of plants with fully sequenced genomes will identify post-translationally modified proteins. Preliminary evidence indicates that some key proteins involved in cell-wall synthesis are regulated by modifications such as phosphorylation. Finally, comprehensive proteomic analysis coupled with imaging analysis is required to identify proteins located in such compartments as the cell wall, nuclear membrane, Golgi, and endoplasmic reticulum.

Cellular System Capabilities

These capabilities are envisioned to help investigators understand how complicated microbial communities in soils respond to various cropping regimes by developing baseline analyses of species composition and abundance in suitable experimental plots. Also needed are cost-effective tools (e.g., diagnostic DNA chips) for monitoring these and other plots over long time periods and for integrating resulting data into a view of how biomass cropping alters soil ecology.

In addition to analyzing soil microbial communities, acquisition and analysis of transcriptomic and proteomic data from model and crop plants will be a powerful tool for assigning probable function to genes implicated in cell-wall synthesis. As a function for each enzyme emerges, the cellular systems capability through imaging and dynamical analyses will be used to develop a systems model to incorporate the wall's biophysical aspects with structural properties and knowledge of the functions of proteins involved in wall synthesis. This model will facilitate the rational development of feedstock species based on "design principles," in which the wall's chemical composition and structure are optimized while, at the same time, not compromising maximal plant productivity. The cellular systems' analytical and modeling resource and proteomic capabilities will carry out a systematic approach for identifying plant biomarkers to guide protein production and the generation of specific molecular tags utilizing protein production capabilities.

The long-term vision is to make available in situ images of living plant cell walls that show all key molecular processes occurring in real time. Studies will cover the full life cycle of cell-wall formation, maturation, transformation, dehydration, and processing into simple feedstocks for conversion into ethanol or production of other fuels and chemicals. The understanding obtained through research that uses such imaging is expected to result in quantitative, predictive modeling as a guide to developing advanced feedstocks and processing them into fuels. These tools will take advantage of various chemically specific imaging tags created with protein production resources.

DOE Joint Genome Institute

DOE JGI will sequence, assemble, and annotate the gene space or entire genomes of model and emerging bioenergy crops and will resequence additional cultivars or ecotypes for marker discovery. Identification of genes that control cell-wall polysaccharide synthesis or modifications in biomass species and the development of tools such as gene chips will depend on the availability of nucleotide sequences. Micro-RNAs are expected to play a role in expression control of many relevant genes, so DNA sequencing must be deep enough to identify them in these species. JGI also will sequence community genomes, including those of the rhizosphere associated with key proposed energy crops and trees.

Other Needed Capabilities

GTL capabilities as currently envisioned do not encompass all resources needed for development of energy crops. To accomplish objectives associated with the energy mission, the current vision should be expanded to include the three capabilities described below.

Transformation Services

Substantial capabilities will be needed to carry out transformation of model and applied species important to feedstock development on behalf of the R&D community and to warehouse genetic resources for these species.

Chemical Phenotyping Services

Feedstock improvement for increased yield and processing will require manipulation and assessment of multiple genes in candidate organisms or model systems. New tools are needed to detect, quantify, and compare changes in cell-wall composition and 3D architecture in primary and secondary walls during assembly and before and during processing. The tools would facilitate feedstock development for improved performance in biomass conversion. Detailed analysis of cell-wall composition by 2D nuclear magnetic resonance (NMR) imaging or other methods is beyond the scope of routine processing in most research laboratories. Capabilities for such analyses would broadly facilitate research.

Synthetic Carbohydrate Chemistry

Analyses of cell walls in their native state and as a function of treatment and processing require advanced synthetic carbohydrate chemistry capabilities to create standards and models. NMR and mass spectroscopy facilities will be necessary to support the synthetic chemistry work.

DECODING THE DNA OF SOYBEAN—A SOURCE OF BIODIESEL

DOE and the U.S. Department of Agriculture will support genome sequencing of the soybean as the first project in an agreement to share resources and coordinate studies of plant and microbial genomics. The soybean, *Glycine max,* is the world's most valuable legume crop and the principal U.S. source of biodiesel, a renewable alternative fuel. Diesel engines inherently are more thermodynamically efficient than combustion engines. Biodiesel has the highest energy content of any alternative fuel and is significantly more environmentally friendly than comparable petroleum-based fuels, since it degrades rapidly in the environment. It also burns more cleanly than conventional fuels, releasing only half the pollutants and reducing the production of carcinogenic compounds by more than 80%. Sequencing will take place at the DOE Joint Genome Institute, supported by DOE's Office of Science. The soybean genome is about 1.1 billion base pairs in size, less than half the size of maize and human genomes.

OTHER BIOFUEL OPPORTUNITIES: DEVELOPMENT OF HIGH-PRODUCTIVITY BIODIESEL CROPS

To maximize solar energy use and storage, an ideal biomass crop would carry out photosynthesis at the theoretical maximum every day of the year and would store fixed carbon in a directly useful and easily harvestable form, even without mineral nutrients. In carrying out this process, many plants accumulate large amounts of oils or waxes in such specialized storage tissues as seeds or mesocarp tissues. In some cases, oil accounts for more than half the dry weight of these tissues. Most oilseed species, however, are not as productive as others such as maize that store primarily starch rather than oil. In theory, obtaining higher yields of oil should be possible by genetically modifying already highly productive species (e.g., sugar beet, potato, and maize) so they will accumulate oils or waxes instead of carbohydrates in their storage and vegetative tissues.

Other plants produced by genetic modifications of developmental processes accumulate oil in their roots (Ogas et al. 1997), suggesting the feasibility of this strategy. Acetyl-CoA carboxylase is the first committed step in fatty-acid biosynthesis in potato tubers, which accumulate starch almost exclusively. Overexpression of this enzyme resulted in a fivefold increase in triacylglycerol accumulation and provided an example of oil deposition in a carbohydrate-storage tissue (Klaus et al. 2004). However, understanding how to reprogram plant cells to store oil rather than carbohydrate is very challenging. It requires not only large-scale changes in the complement of expressed metabolic enzymes but also changes in cellular structure typically associated with cell identity.

Because recovery of oil from plants is technically simple and efficient, processing oil-accumulating plant material on the farm may be possible so mineral nutrients and soil adjuvants can be retained at the farm for return to the soil (see sidebar, Decoding the DNA of Soybean—A Source of Biodiesel, p. 77). In principle, oil could be recovered during harvest by inexpensive screw-press technology, greatly reducing transport and processing costs and enhancing sustainability.

The lipids that comprise most plant oils are highly reduced forms of carbon and therefore represent the most energy-dense plant-storage compounds. Derivatives such as biodiesel can be produced easily by inexpensive and efficient conversion of plant-derived oil to fatty acid methyl esters. Production of high-yielding, oil-accumulating plants could result in a new biofuel source that would reduce loss of carbon associated with fermenting sugars to ethanol as well as costs associated with converting lignocellulosic feedstocks to liquid fuels. The plant that most nearly meets this description is oil palm (see sidebar, Oil Palm, this page).

OIL PALM: AN IMPORTANT BIOFUEL PLANT

Oil palm, *Elaeis guineensis* Jacq., is a tropical tree species that produces bunches of oil-rich fruit resembling avocados. The plants grow in lowlands of the humid tropics (15°N to 15°S), where rainfall of 1800 to 5000 mm is evenly distributed throughout the year. Oil palms mature slowly but, once established, yield as much as 10.6 tonnes of oil per hectare (ha) per year, although the average is less than half that amount. They begin to bear fruit after about 3 years and remain in use for some 25 years, so annual maintenance costs are low.

Palm oil from the mesocarp, which contains 45 to 55% oil, is similar in composition to soy oil. Oil from palm kernels, extracted from the endosperm, contains about 50% oil rich in medium-chain fatty acids and well suited for biodiesel applications. Malaysia is the major source of these oils, producing 13.4 billion pounds of mesocarp oil and 3.5 billion pounds of kernel oil from about 3.8 million ha. Plants are harvested by hand, but typical planting density is only 150 plants/ha, and labor costs are not a major factor in production. Because of the relatively straightforward conversion of palm oils to diesel and food uses, palm acreage in the tropics is expected to expand significantly. [Picture source: C. Somerville, Stanford University]

Some questions to be addressed by basic research:

- How is cell identity controlled? More specifically, what high-level, regulatory genetic controls program a cell to express genes involved in lipid synthesis and accumulation?
- What regulates carbon flux from photosynthetic-source leaves to storage organs?
- What regulates metabolic partitioning of carbon among proteins, starch, and oil production?
- What species are most promising for conversion to oil accumulation? For instance, is oil potato a better candidate than oil beet or very high oil maize, or should plant stems be engineered to deposit extremely thick wax-rich cuticles?
- What are the opportunities for developing new high-yielding oilseed species tailored to energy production by uncoupling oil accumulation from seed carbohydrate or protein accumulation?
- What is the impact of high lipid levels on plant cells? Most cellular mechanisms operate in aqueous environments.

CITED REFERENCES

Albrecht, W. A. (1938). "Loss of Soil Organic Matter and Its Restoration," pp. 347–60 in *Soils and Men*, U.S. Department of Agriculture.

Brown, G. R., et al. (2003). "Identification of Quantitative Trait Loci Influencing Wood Property Traits in Loblolly Pine (*Pinus taeda L.*) III. QTL Verification and Candidate Gene Mapping," *Genetics 164,* 1537–46.

Brunner, A. M., V. Busov, & Strauss, S. H. (2004). "The Poplar Genome Sequence: Functional Genomics in a Keystone Plant Species," *Trends Plant Sci. 9,* 49–56.

Busov, V. B., et al. (2005). "Insertional Mutagenesis in *Populus*: Relevance and Feasibility," *Tree Genet. Genomes 1,* 135–42.

Christiansen, P., et al. (2005). "A Rapid and Efficient Transformation Protocol for the Grass *Brachypodium distachyon*," *Plant Cell Rep. 23,* 751–58.

Draper, J., et al. (2001). "*Brachypodium distachyon*: A New Model System for Functional Genomics in Grasses," *Plant Phys. 127,* 153 9–55.

EERE. (2003) (Review Draft). *Biomass Program Multi-Year Technical Plan*, U.S. DOE, Energy Efficiency and Renewable Energy (http://devafdc.nrel.gov/biogeneral/Program_Review/MYTP.pdf).

Fitter, A. H., et al. (2005). "Biodiversity and Ecosystem Function in Soil," *Funct. Ecol. 19* (3), 369–77.

Follett, R. F. (2001). "Soil Management Concepts and Carbon Sequestration in Cropland Soils," *Soil Tillage Res. 61,* 77–92.

Groover, A., et al. 2004. "Gene and Enhancer Trap Tagging of Vascular-Expressed Genes in Poplar Trees," *Plant Physiol. (134),* 1742–51.

Handelsman, J., et al. (1998). "Molecular Biological Access to the Chemistry of Unknown Soil Microbes: A New Frontier for Natural Products," *Chem. Biol. 5,* R245–49.

Klaus, D., et al. (2004). "Increased Fatty Acid Production in Potato by Engineering of Acetyl-CoA Carboxylase," *Planta 219,* 389–96.

Koonin, S. E. (2006). "Getting Serious About Biofuels," *Science 311,* 435.

Maskina, M. S., et al. (1993). "Residual Effects on No-Till Crop Residues on Corn Yield and Nitrogen Uptake," *Soil Sci. Soc. Am. j 57,* 1555–60.

Neale, D. B., and O. Savolainen. 2004. "Association Genetics of Complex Traits in Conifers," *Trends Plant Sci. 9,* 325–30.

Ogas, J., et al. (1997). "Gibberellin Mediates the Transition from Embryonic to Vegetative Cell Identity During Germination of the *Arabidopsis pickle* Mutant," *Science, 277,* 91–94.

Perlack, R. D., et al. (2005). *Biomass as Feedstock for a Bioenergy and Biop roducts Industry: The Technical Feasibility of a Billion-Ton Annual Supply*, DOE/ GO-102005-2135, Oak Ridge National Laboratory, Oak Ridge, Tennessee (http://feedstockreview.ornl.gov/pdf/billion_ton_vision.pdf).

Rafalski, A. (2002). "Applications of Single Nucleotide Polymorphisms in Crop Genetics," *Curr. Opin. Plant Biol. 5,* 94–100.

Remington, D. L., et al. (2001). "Structure of Linkage Disequilibrium and Phenotypic Associations in the Maize Genome," *Proc. Natl. Acad. Sci. USA 98,* 11479–84.

Sterky, F., et al. (2004). "A *Populus* Expressed Sequence Tag Resource for Plant Functional Genomics," *Proc. Natl. Acad. Sci. USA* (www.pnas.org/ cig/reprint/101/3 8/13951).

Tuskan, G. A., et al. In press (2006). "The Genome of Black Cottonwood, *Populus trichocarpa*," *Science*.

Tuskan, G. A., DiFazio, S. P. & Teichmann, T. (2004). "Poplar Genomics is Getting Popular: The Impact of the Poplar Genome Project on Tree Research," *Plant Biol. 6,* 2–4.

Vogel J. P., et al. (2006). "*Agrobacterium*-Mediated Transformation and Inbred Line Development in the Model Grass *Brachypodium distachyon*," *Plant Cell Tiss. Org. Cult.* (www.springerlink.com, doi: 10.1007/s11240-005-9023-9).

Wilhelm, W. W., et al. (2004). "Crop and Soil Productivity Response to Corn Residue Removal: A Literature Review," *Agron. J. 96,* 1–17.

FOR FURTHER READING

American Society of Plant Biologists. (2005). "President Bush Cites Importance of Plant-Based Biofuels," *ASPB News* 32 *(4),* 16–17.

Bar-Joseph, Z., et al. (2003). "Computational Discovery of Gene Modules and Regulatory Networks," *Nat. Biotechnol. 21,* 1337–42.

Brown, G. R., et al. (2004). "Nucleotide Diversity and Linkage Disequilibrium in Candidate Genes Influencing Wood Formation in Loblolly Pine," *Proc. Natl. Acad. Sci. 101(42),* 15255–260.

Brummer, E. C. (1999). "Capturing Heterosis in Forage Crop Cultivar Development," *Crop Sci. 39,* 943–54.

Busov, V. B., et al. (2005). "Genetic Transformation: A Powerful Tool for Dissection of Adaptive Traits in Trees," *New Phytol. 167,* 219–28.

Cardon, L. R., & Bell, J. L. (2001). "Association Study Designs for Complex Diseases," *Nat. Rev. Genet. 2,* 91–99.

Del Grosso, S. J., et al. (2005). "DAYCENT Model Analysis of Past and Contemporary Soil N_2O and Net Greenhouse Gas Flux for Major Crops in the USA," *Soil Tillage Res. 83,* 9–24.

DiFazio, S. P. (2005). "A Pioneer Perspective on Adaptation," *New Phytol. 165,* 661–64.

Flint-Garcia, S. A., Thornsberry, J. M. & Buckler, E. S. IV. (2003). "Structure of Linkage Disequilibrium in Plants," *Annu. Rev. Plant Biol. 54,* 357–74.

Follett, R. F., et al. (2005). "Research and Implementation Needs to Mitigate Greenhouse Gas Emissions from Agriculture in the USA," *Soil Tillage Res. 83,* 159–66.

Follett, R. F., Castellanos, J. Z. & Buenger. E. D. (2005). "Carbon Dynamics and Sequestration in an Irrigated Vertisol in Central Mexico," *Soil Tillage Res. 83,* 148–58.

Franzluebbers, A. J. (2005). "Soil Organic Carbon Sequestration and Agricultural Greenhouse Gas Emissions in the Southeastern USA," *Soil Tillage Res. 83,* 120–47.

Franzluebbers, A. J., & Follett, R. F. (2005). "Greenhouse Gas Contributions and Mitigation Potential in Agricultural Regions of North America: Introduction," *Soil Tillage Res. 83,* 1–8.

Gregorich, E. G., et al. (2005). "Greenhouse Gas Contributions of Agricultural Soils and Potential Mitigation Practices in Eastern Canada," *Soil Tillage Res. 83,* 53–72.

Harbison, C. T., et al. (2004). "Transcriptional Regulatory Code of a Eukaryotic Genome," *Nature 431*, 99–104.

Hooker, B. A., et al. (2005). "Long-Term Effects of Tillage and Corn Stalk Return on Soil Carbon Dynamics," *Soil Sci. Soc. Am. J. 69*, 188–96.

Hu, W.-J., et al. (1999). "Repression of Lignin Biosynthesis Promotes Cellulose Accumulation and Growth in Transgenic Trees," *Nat. Biotechnol. 17*, 808–12.

Huggins, D. R., et al. (1998). "Soil Organic C in the Tallgrass Prairie- Derived Region of the Cornbelt: Effects of Long-Term Crop Management," *Soil Tillage Res. 47*, 219–34.

Jawson, M. D., et al. (2005). "GRACEnet: Greenhouse Gas Reduction Through Agricultural Carbon Enhancement Network," *Soil Tillage Res. 83*, 167–72.

Johnson, J. M. F., et al. (2005). "Greenhouse Gas Contributions and Mitigation Potential of Agriculture in the Central USA," *Soil Tillage Res. 83*, 73–94.

Johnson, J. M. F., Allmaras, R. R. & Reicosky. D. C. (2006). "Estimating Source Carbon from Crop Residues, Roots and Rhizodeposits Using the National Grain-Yield Database," *Agron. J.* (in press).

Lal, R. (2004). "Is Crop Residue a Waste?" *j Soil Water Conserv. 59*, 136–39.

Larson, W. E., et al.(1972). "Effects of Increasing Amounts of Organic Residues on Continuous Corn. II. Organic Carbon, Nitrogen, Phosphorus, and Sulfur," *Agron. J. 64*, 204–8.

Liebig, M. A., et al. (2005). "Greenhouse Gas Contributions and Mitigation Potential of Agricultural Practices in Northwestern USA and Western Canada," *Soil Tillage Res. 83*, 25–52.

Linden, D. R., Clapp, C. E. & R. H. Dowdy. (2000). "Long-Term Corn Grain and Stover Yields as a Function of Tillage and Residue Removal in East Central Minnesota," *Soil Tillage Res. 56,* 16 7–74.

Mann, L. K., Tolbert, V. R. & J. Cushman. (2002). "Potential Environmental Effects of Corn (*Zea mays* L.) Stover Removal with Emphasis on Soil Organic Matter and Erosion: A Review," *Agric. Ecosyst. Environ. 89*, 149–66.

Martens, D. A., et al. (2005). "Atmospheric Carbon Mitigation Potential of Agricultural Management in the Southwestern USA," *Soil Tillage Res. 83*, 95–119.

Mukherjee, S., et al. (2004). "Rapid Analysis of the DNA-Binding Specificities of Transcription Factors with DNA Microarrays," *Nat. Genet. 36,* 1331–39.

Nelson, R. G. (2002). "Resource Assessment and Removal Analysis for Corn Stover and Wheat Straw in the Eastern and Midwest United States— Rainfall and Wind-Induced Soil Erosion Methodology," *Biomass Bioenergy 22,* 349–63.

Nordborg, M., et al. (2002). "The Extent of Linkage Disequilibrium in *Arabidopsis thaliana,*" *Nat. Genet. 30*, 190–93.

Strauss, S. H., et al. (2004). "Ten Lessons from 15 Years of Transgenic *Populus* Research," *Forestry 77*, 455–65.

Thornsberry, J. M., et al. (2001). "*Dwarf8* Polymorphisms Associate with Variation in Flowering Time," *Nat. Genet. 28,* 286–89.

Tong, A. H. Y., et al. (2004). "Global Mapping of the Yeast Genetic Interaction Network," *Science 303,* 808–13.

U.S. DOE Office of Science. 2005. "Making Bioethanol Cost Competitive," DOE Genomics:GTL Program, Systems Biology for Energy and Environment (www.doegenomestolife.org/pubs/Bioethanol10_27_05_8.5.pdf).

Venter, J. C., et al. 2004. "Environmental Genome Shotgun Sequencing of the Sargasso Sea," *Science, 304*, 66–74.

Vogel, K. P. (2004). "Switchgrass," pp. 56 1–88 in *Warm-Season (C4) Grasses*, ed. L. E. Moser, L. Sollenberger, and B. Burson, ASA-CSSA-SSSA Monograph No. 45, Madison, Wis.

Vogel, K. P. (2000). "Improving Warm-Season Grasses Using Selection, Breeding, and Biotechnology," pp. 83–106 in *Native Warm-Season Grasses: Research Trends and Issues*, ed. K. J. Moore and B. Anderson, Crop Science Special Publication No. 30, Crop Science Society of America and American Society of Agronomy, Madison, Wis.

Vogel, K. P., & Burson. B. (2004). "Breeding and Genetics," pp. 51–96 in *Warm-Season (C4) Grasses*, ed. L. E. Moser, L. Sollenberger, and B. Burson, ASA-CSSA-SSSA Monograph No. 45, Madison, Wis.

Vogel, K. P., & Jung, H. G. (2001). "Genetic Modification of Herbaceous Plants for Feed and Fuel," *Crit. Rev. Plant Sci., 20*, 15–49.

Walsh, M. E., et al. (2003). "Bioenergy Crop Production in the United States," *Environ. Resour. Econ. 24(4)*, 313–33.

Chapter 5

DECONSTRUCTING FEEDSTOCKS TO SUGARS

United States Department of Energy

This chapter describing the challenges of deconstructing cellulosic biomass to ethanol is critically linked to both the feedstock and fermentation areas. Lignocellulose recalcitrance to bioprocessing will remain the core problem and will be the limiting factor in creating an economy based on lignocellulosic ethanol production. Understanding biomass recalcitrance will help to drive crop design. Knowledge about feedstock breakdown mechanisms and products will drive fermentation and consolidation strategies, which ultimately will include consolidated bioprocessing (CBP). CBP incorporates the traits for deconstruction and fermentation of sugars to ethanol into a single microbe or culture. The previous chapter describes how tools of modern genomics-based systems biology can provide tremendous opportunities to engineer energy-plant genomes for new varieties. Those engineered plants will grow more efficiently while also producing optimal polysaccharide compositions for deconstruction to sugars and fermentation to ethanol and other products. Further advancements in plant engineering also can generate new energy crops and trees harboring polysaccharide storage structures (principally in the cell walls) that are *designed* for deconstruction. This achievement will be an important outcome of scientific research needed to optimize deconstruction of native cell walls in such crop residues as corn stover and wheat straw and such energy crops as switchgrass and poplar.

Once we understand more about the chemistry and ultrastructure of cell walls, improved thermochemical and biological means can be used to disassemble them. For example, this report will show that new and improved (existing) enzymes capable of depolymerizing cellulose in cell-wall micro-fibrils can significantly reduce the cost of deriving sugars from biomass—a critical factor in lowering the overall cost of cellulosic ethanol and making it cost-competitive with gasoline. Currently, the structure of cellulose itself is not adequately understood. For example, we do not know the actual faces of the cellulose crystal to which cellulases bind, and surface interactions among experimentally produced enzymes and substrates are not understood. In addition, such new findings about lignin as its synthesis, modification, and depolymerization are needed to develop conversion processes that are less energy intensive.

Sugars produced in optimal cell-wall deconstruction will be used by microbes specializing in converting (fermenting) these compounds to ethanol. In the near term, biomass conversion will produce some quantities of fermentation inhibitors that can include acetate, polyphenolics (aromatic compounds derived from lignin breakdown), and cellobiose. Even high concentrations of sugars from cell-wall deconstruction are inhibitory to many ethanologens. Thus the chapter, Sugar Fermentation to Ethanol, p. 119, describes research to develop processes at the cellular and genomic levels (e.g., evolutionary engineering to modify microbial strains that tolerate elevated levels of toxins or sugars). Also, genomic and bioinformatic tools can assist in the design of new metabolic pathways that permit efficient simultaneous fermentation of mixed sugars.

Looking ahead to longer-term improvements in biorefinery productivity, this report proposes that new organisms consolidating traits for both deconstruction and fermentation should be the subject of considerable research. For example, new ethanologen strains are needed that also are able to produce mixtures of highly competent cellulases and other hydrolytic enzymes. The application of advanced techniques in genetics, molecular biology, high-throughput screening, imaging, and mathematical modeling will accelerate the pace at which viable processing can be accomplished for energy security.

DETERMINING FUNDAMENTAL PHYSICAL AND CHEMICAL FACTORS IN THE RECALCITRANCE OF LIGNOCELLULOSIC BIOMASS TO PROCESSING

Lignocellulosic biomass is a complex structure with crystalline cellulose, hydrated hemicellulose, and lignin as major components. To date, the best enzyme cocktails proposed for saccharification of this material are synergistic mixtures of enzymes with defined activities, primarily those that degrade cellulose. The substrate's heterogeneous nature and enzyme cocktails' complexity suggest that traditional studies of bulk properties will not provide the detailed understanding required for knowledge-based advancements in this field.

The question that must be answered is, How will structural and chemical details of enzyme substrate–binding sites affect enzyme adsorption and reaction rates? From cellulase kinetics alone, we never would be able to answer that question. We must be able to improve cellulases using an informational approach, considering that 20 years of "mixing and testing" individual cellulase proteins has yielded only modest progress toward an improved cellulase system. Enzyme kinetics always has been a measure of "ensemble average" results from experiments, and thus sugar-release values from biomass never can deliver information about the reactive site of the individual enzyme. Because of this dilemma, we do not know if cellulose recalcitrance is due to enzyme inadequacy, enzyme-substrate mismatch, or both. While in many cases we can perform bulk compositional analyses, molecular and structural correlations that are key to processability remain a challenge. The science and technology for these analyses must come from frontier capabilities in many disciplines in the physical, biological, and computational sciences [see sidebar, Image Analysis of Bioenergy Plant Cell Surfaces at the OBP Biomass Surface Characterization Lab (BSCL), p. 40].

Such enzymes as cellulases, hemicellulases, and other glycosyl hydrolases (GH) synthesized by fungi and bacteria work synergistically to degrade structural polysaccharides in biomass. These enzyme systems, however, are as complex as the plant cell-wall substrates they attack. For example, commercial cellulase preparations are mixtures of several types of GH, each with distinctly different substrate specificities (cellulose and xylan) and action patterns (exoenzymes acting from the chain ends, endoenzymes cleaving within the chain, and GH cleaving side-chain branches, for example, arabinose from arabinoxylans and xylose from xyloglucans). Optimization of these enzymes will require a more detailed understanding of their regulation and activity as a tightly controlled, highly organized system.

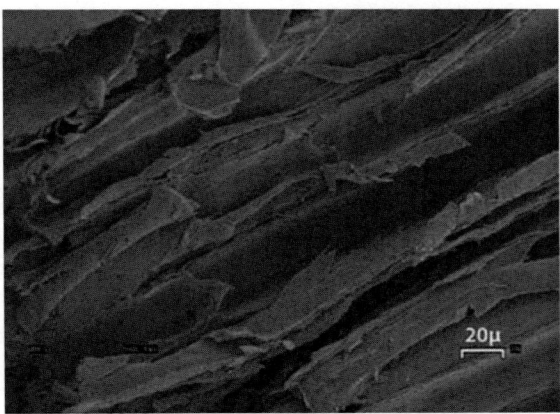

(a) Enzyme Hydrolysis Only.

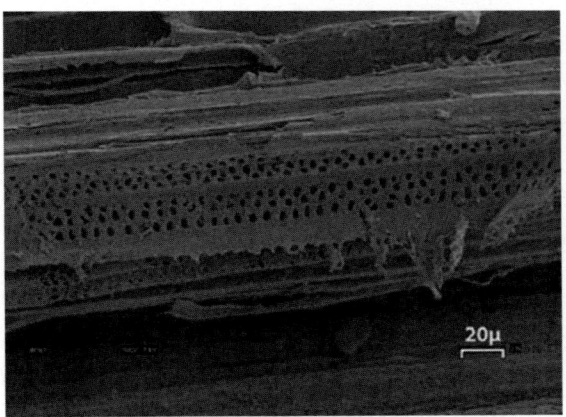

(b) Enzyme Hydrolysis Following Pretreatment.

Source: Images and conditions from unpublished work of M. Zeng, N. Mosier, C. Huang, D. Sherman, and M. Ladisch, 2006.

Figure 1. Scanning Electron Microscopy.
(a) A corn stover particle shows a smooth surface with a few micron-sized pores after enzyme hydrolysis converted 11% of cellulose to glucose in 3 h. (b) This corn stover particle has many more pores. It was pretreated in water at 190°C for 15 min and hydrolyzed by enzymes at 50°C for 3 h, resulting in 40% cellulose conversion to glucose.
The results illustrate that pretreatment changes lignocellulosic-structure susceptibility to attack by enzymes. Higher resolution in future imaging techniques will facilitate a deeper understanding of underlying molecular mechanisms.

Tools must be developed first to analyze biomass at the levels and specificities needed to describe, understand, predict, and control its behaviors. These include molecular-level and nanoscale understanding of critical cell-wall physical and chemical properties, how they are formed in the plant, how they function, and how certain structural features inhibit or facilitate enzymatic interactions and subsequent saccharification. Due to the dynamic and interconnected nature of these conversion processes, we must be able to monitor them as they occur—from native structural biomass to free sugars. Critical interactions and reactions occur at many scales of time and length throughout multiple stages of intricately interlinked processing. All biomass structural and chemical properties must be analyzed in the native state, in various stages of dehydration and processing, and as the material is deconstructed into complex mixtures of reaction intermediates. These analyses must support research for biomass optimization for processing and ethanol yield, as well as investigations of robust growth and tolerance in the cultivar. Pretreatment optimization will both maximize fermentable sugar yield while minimizing inhibitors and other deleterious factors in subsequent steps.

This problem further involves structure and function at both the plant and microbial levels; their interactions; and the functions of proteins, enzymes, and living systems. They are studied under a very wide range of conditions, from pretreatment temperatures and chemistry to the enzymatic break-down of complex cellulosic microfibrils and other polymers into sugars. To understand reaction pathways, we must be able to discern intermediate digestion and reaction products—physical structures as they degrade and chemical moieties as they are transformed and react. All these capabilities will enable the ultimate consolidation of all steps into one process.

An important and necessary departure from traditional approaches stems from our appreciation that biological systems are incredibly complex and interconnected. Cellular systems of plants and biomass-degrading microbes consist of highly coupled subsystems. In these living systems, each component (molecules and macromolecules) affects and is affected by other components. Resolution of research issues described in this section will advance biomass pretreatment and hydrolysis by developing a systems-level, predictive, and quantitative new understanding. Such understanding will facilitate rational design of biomass-conversion systems, inform the design of a new generation of energy crops, and accelerate adoption of innovative biorefinery technologies.

Research Goals

Biological, mathematical, imaging, and other analytical tools must be developed and applied to identify and quantify the relative importance of potentially limiting (or facilitating) physical and chemical factors in bioconversion. All these methods will be used to characterize the effects of dehydration, heating, acidity, and cosolvents, among other biomass treatments. Detailed studies using these tools will help us better understand mechanisms of action and apply various pretreatments and enzyme systems to enhance lignocellulose bioconversion to ethanol. Such topics are discussed below.

Measurement of Biomass Properties

The organization of polymer components in biomass structures needs to be noninvasively imaged in 3D to elucidate the material's organization. Various experimental approaches including X-ray and neutron diffraction of crystals and drawn fibers, enhanced atomic force microscopy (AFM), and biochemical probe–enabled and chemically specific imaging will be used to characterize the different forms of lignocellulose in both "native" and treated (stored and processed) biomass cell walls.

Analytical methods [e.g., nuclear magnetic resonance (NMR), Fourier transform infrared (FTIR), and Raman] should provide information on chemical moieties, chemical bonds, and conformation of wall polymers. Surfaces that are enzyme-binding sites should be characterized by such tools as AFM, scanning electron microscopy, transmission electron microscopy, and electron spectroscopy for chemical analysis (see Figure. 1. Scanning Electron Microscopy, this page).

This information can be correlated using statistical-analysis packages based on principal components to reliably relate changes in lignin carbohydrate complexes (LCC) and lignin profiles, degrees of cellulose dehydration, and pretreatment chemistries to the abilities of hydrolytic enzymes to convert the resultant material biomass.

Models for Direct Enzymatic Interactions, Action

Mechanistic models of enzyme-system substrate relationships, based on new knowledge about plant cell-wall architecture and enzyme structure and function, need to be developed. The action of individual and combinations of enzymes on native and model substrates should be examined using single-molecule spectroscopy and imaging.

Mechanistic characterization of relevant enzymes, with attention to combined chemical and biological conversion processes of cell-wall materials, then will support models for interaction. Finally, new computational models for microfibrils and cell walls using improved codes and petascale leadership-class computers will provide the needed fidelity to support reaction-pathway calculations.

Technical Milestones

Within 5 years

- Using integrated analytical and mathematical methods, quantitatively analyze effects of selected biomass properties and pretreatment chemistries on enzymatic hydrolysis.
- Identify principles, genes, and controlling factors that influence biomass makeup, assembly, and processability, in conjunction with plant design.
- Perfect and extend existing spectroscopic, surface, and imaging characterization methods for biomass.
- Develop methods for monitoring the progress of physical modification during dehydration under various conditions of storage or pretreatment at elevated temperatures. Develop improved protocols to be tested in field trials.
- Define the primary structure of lignins and LCCs and the shape, dimensions, and heterogeneity of their domains.

- Develop models connecting lignocellulosic properties with deconstruction and hydrolysis processes.
- Definitively characterize the detailed organizational structures of principal types of plant cellulose and their relative energies and interrelationships.
  Understand the energetics of different cellulose forms and ways in which these structures give rise to characteristics of the next organization level in the mesoscopic range of sizes, and
  Determine how microfibrils interact with other principal structural components (lignins and noncellulosic polysaccharides).

Within 10 years

- Develop an understanding of the physicochemical basis of polysaccharide interactions with water. This information can be used to rationally design genomic variations for cell-wall polysaccharide composites with more desirable properties.
- Apply new biological and chemical tools to alter LCC and other relevant properties of selected biomass species.
- Subject these altered biomass species and controls to various pretreatment chemistries and temperatures.
- Hydrolyze the resulting pretreated biomass with various specific enzymes and correlate spectroscopic, surface, and image data with hydrolysis of structural carbohydrates.
- Quantitatively relate key biomass and pretreatment properties with hydrolysis yields for different biomass species.
- Improve enzymes based on the knowledge of substrate-imposed limitations and an understanding of optimized cell walls.
- Develop molecular models for enzyme-substrate structure-function relationships.

Within 15 years

- Develop mathematical models of cell walls and the transport of water, pretreatment chemicals, and enzymes through the walls. Such models should help in designing cell walls that can be digested more efficiently by developed enzyme systems. Designs optimized by modeling can then help guide genomic engineering for plant and enzyme optimization.
- Establish the foundation for integrated biomass processing by predicting and then verifying overall hydrolysis yields for native and modified biomass species using different pretreatment chemistries and severities (time, temperature, acidity) and a range of hydrolytic enzymes.
- Perform rational engineering of enzymes to specific requirements based on new understandings about enzymes and their mechanisms of action and substrate cellulose-hemicellulose-lignin complexes.

The Role of GTL and OBP Facilities and Capabilities

The program outlined most likely will need all facility resources at one point or another.

Protein Production
Protein production capabilities will be used for enzymes and tags (biomarkers).

Molecular Machine and Cellular System Analysis
These capabilities can image and diagnose interactions between enzymes and biomass substrates. In addition, BSCL, recently established at NREL, certainly will prove essential. Imaging facilities would be used to correlate digestibility to biomass ultrastructure. The sidebar, Image Analysis of Bioenergy Plant Cell Surfaces at the OBP Biomass Surface Characterization Lab, p. 40, illustrates the cell wall's highly coupled nature and how each component affects and is affected by other components, thereby highlighting the need to study the intact system.

DOE Joint Genome Institute
DOE JGI will map and identify genes concerned with deconstruction barriers or facilitating factors.

Crosscutting Tools, Technologies, and Science

- Better characterization of cell-wall polysaccharide association profiles in hydrated and dehydrated states using spectroscopic and ultramicroscopic methods.
- Larger-scale molecular mechanics modeling to explore the association of cell-wall polymers with each other and with water at various length scales.
- Robust and rapid methods for quantifying cell-wall polymers and inter-polymer linkages.
- Genomic and chemical tools to identify genes involved in rate-limiting LCC linkages.
- New biomass variants with enhanced accessibility for structural carbohydrate–digesting enzymes.
- New methods for spectroscopic characterization of cell-wall polysaccharides in the living hydrated state. Most past work has been based on investigation of isolated dehydrated samples and very little on cell-wall constituents in their native state.
- Coupled structure and processing models of cell walls and components.
- Data-analysis software to identify and quantitatively describe the relative importance of different factors governing response to pretreatment and enzymatic hydrolysis of biomass.

DEVELOPING BETTER ENZYMATIC SYSTEMS FOR BIOLOGICAL PRETREATMENT: LIGNINASES AND HEMICELLULASES

The most efficient conditions for simultaneous saccharification and fermentation of lignocellulosic biomass will be those in which transformations now performed during and following thermochemical pretreatment are seamlessly integrated into the overall process. Effective enzymatic breakdown of LCC, to expose cellulose to enzyme action, represents an important step toward this objective.

Lignocellulose, an extremely complex and widely varying nanoscale composite, is well designed to resist attack. Ligninases and hemicellulases, for which few examples are known, are inadequately understood. Technically, understanding and optimizing these enzymes will enable the ultimate goal of consolidating pretreatment and saccharification. Consequently, research is needed to identify, characterize, improve, and economically produce the most effective enzyme systems for biomass preconditioning. These enzymes would be applied before or after traditional pretreatment to minimize and, eventually, replace thermochemical processes, thus lessening the effects of overall pretreatment severity at the macromolecular level and simplifying processing. To support improvements, research must be focused on identifying more enzymes of this class and characterizing their principles of action. A crucial component for ligninase identification will be genomic, functional genomic, and proteomic comparisons of white rot fungi known or presumed to express such activities (see sidebar, White Rot Fungus, p. 93).

The ultimate goal of this research is to produce a recombinant ligninasehemicellulase microbial system with enhanced catalytic activity and stability, industrialized for biorefinery operations.

Research Goals

The chemical and structural nature of native lignin is poorly understood. Developing robust ligninases requires a foundation of knowledge about the range of lignin chemistries and lignin associations that inhibit or facilitate freeing cellulose and hemicellulose for saccharification (see sidebar, Lignification: Random vs Template Directed, p. 94). Similarly, despite the importance of hemicelluloses and hemicellulases for saccharification, both are poorly understood. Due to the complex structure and compositional diversity of lignin-hemicellulose and -cellulose substrates that must be contemplated for feedstock biomass, a variety of enzymatic activities will be needed to catalyze deconstruction for conversion to monosaccharides (see Figure. 2. Complex Mixture of Enzymes for Degrading Hemicelluloses, p. 96).

Hemicelluloses and Hemicellulases

After cellulose, hemicelluloses (xylan polymers) are the next most abundant polysaccharides in native biomass feedstocks (see Table 1. Cellulose, Hemicellulose, and Lignin Content in Various Sources of Biomass, p. 93). Although considerable research has focused on converting biomass cellulose to fermentable glucose, less has been done on bioconverting other plant cell-wall components. A better understanding of the complex structure and composition of this polysaccharide group, which will differ in type and abundance among different feedstocks, will help to identify and optimize the mechanistic basis of required enzymatic activities. Structural information and mechanistic models must be developed to pinpoint "bottlenecks" in hemicellulose bioconversion.

Lignin and Ligninases

Despite its critical importance, the enzymatic basis of lignin depolymerization in vivo has remained elusive, if not controversial. Research during the past 20 years on putative "ligninases," which can both polymerize and depolymerize lignin preparations, has not yielded reliable insights into the mechanisms of lignin cleavage. Thus the possibility of another entirely different class of lignin-degrading enzymes cannot be disregarded, and potential candidates have been identified in white rot fungi. Genes encoding a range of ligninases must be identified, and genomic sequences of white rot fungi that degrade lignin but do not express known peroxidases will be compared with the sequenced *P. chrysosporium* genome. A suitable host organism will be needed to produce heterologous ligninases in sufficient quantities; possible hosts include *E. coli, S. cerevisiae, P. pastoris,* and *A. oryzae*. Heterologous ligninase expression and manipulation have had limited success.

WHITE ROT FUNGUS: GENOME OF KNOWN LIGNIN DEGRADER SEQUENCED

Lignin degradation is the key to making polysaccharide components of cell walls available for breakdown. White rot fungi are the primary degraders of lignin, which is among the most abundant of natural materials on earth and plays a pivotal role in global carbon cycling. These organisms also degrade the cellulose and hemicellulose components of plant cell walls.

To aid in understanding these processes, the DOE Joint Genome Institute sequenced the genome of the white rot fungus *Phanerochaete chrysosporium*. This fungus degrades brown lignin, the protective matrix surrounding cellulose microfibrils of plant cell walls, leaving behind crystalline white cellulose. [D. Martinez et al., "Genome Sequence of the Lignocellulose Degrading Fungus *Phanerochaete chrysosporium* strain RP78," *Nat. Biotechnol.* **22,** 695–700 (2004).]

Analysis of the white rot fungus genome revealed genes encoding oxidases, peroxidases, and other enzymes that contribute to depolymerization of lignin, cellulose, and hemicellulose. Extensive genetic diversity was observed in gene families encoding these enzymes, possibly reflecting that multiple specificities are needed for effective degradation of cell-wall polymers from different plant species. Elucidating the regulation of genes, proteins, and metabolites from this organism and others will enhance understanding of the individual and collective mechanisms of degradative enzymes as well as their interactions with other organisms in their ecosystems. Such advances are necessary for generating the framework to engineer large-scale processes for biomass utilization.

Table 1. Cellulose, Hemicellulose, and Lignin Content in Various Sources of Biomass

Feedstock	Cellulose	Hemicellu.	Lignin
Corn stover	36.4	22.6	16.6
Wheat straw	38.2	24.7	23.4
Rice straw	34.2	24.5	23.4
Switchgrass	31.0	24.4	17.6
Poplar	49.9	20.4	18.1

Source: A. Wiselogel, S. Tyson, and D. Johnson, "Biomass Feedstock Resources and Composition," pp. 105–18 in *Handbook on Bioethanol: Production and Utilization* (Applied Energy Technology Series), ed. C. E. Wyman, Taylor and Francis (1996).

This research would include the following goals:

- Detailed characterization of the range of lignin and hemicellulose struc tures encountered in available biomass. Special attention must be paid to the nature of covalent bonds linking LCC. Model systems will facilitate these analyses.
- Systems biology–style survey and molecular and functional characterization of the diversity and activity of relevant hemicellulases (glycoside hydrolases and esterases), including discovery of activities from such novel sources as soils, the rhizosphere, termite hindgut, and decaying biomass.
- Similar discovery and molecular and functional characterization of new lignin-degrading enzymes and their activities.
- Reliable expression systems and hosts for these enzyme classes.
- Mechanistic characterization of enzymes with attention to properties relevant to combined chemical and biological conversion processes.
- Examination of individual and combination enzyme activities on native and model substrates.
- Development of mechanistic models of enzyme system substrate relationships, based on new knowledge about plant cell-wall architecture.

- Improvement of enzyme catalytic efficiency through protein engineering and of biomass through plant design.
- Coordination with feedstocks and biological-conversion research to inform plant design and to support improved fermentation and eventual incorporation of pretreatment enzymatic processes into consolidated processing schemes. Lignin and hemicellulose investigations will follow similar research paths as described below.

LIGNIFICATION: RANDOM VS TEMPLATE DIRECTED

In terms of energy content, lignins are thought to be the most abundant of all biopolymers. They are composed of p-hydroxyphenylpropanoid units interconnected through 8-O-4, 8-5, 8-8, 8-1, 5-5, and 4-O-5 linkages. Corresponding substructures in the polymer include alkyl aryl ethers, phenylcoumarans, resinols, tetrahydrofuran-spiro-cyclohexadienones, biphenyls, dibenzodioxocins, and diaryl ethers (see Figure. A, at left). The primary precursors themselves—the three monolignols p-coumaryl, coniferyl, and sinapyl alcohols—differ only according to their aromatic methoxy substitution patterns. These monolignols are oxidized enzymatically through single-electron transfer to generate the respective phenoxy radicals. The actual coupling of a monolignol radical with the growing end of a lignin chain, however, may not fall under direct enzymatic control.

Accordingly, many investigators have assumed that lignin primary structures must be "random" or combinatorial as far as sequences of interunit linkages are concerned. More recently, this theory has been reinforced by reports that certain kinds of non-native monolignols can be incorporated into macromolecular lignin structures. Lignins and lignin derivatives exhibit two fundamental characteristics that traditionally have been viewed as evidence in favor of randomness in their configurations: They are both noncrystalline and optically inactive.[1]

Nevertheless, a number of observations are thought by some to point in the opposite direction. The individual molecular components in (nonpolyionic) lignin preparations tend to associate very strongly with one another in a well-defined way. These processes are thought to be governed by vital structural *motifs* derived from corresponding features disposed nonrandomly along the native biopolymer chain. Moreover, dimeric pinoresinol moieties are linked predominantly to the macromolecular lignin chain through at least one of their aromatic C-5 positions.

We do not know whether such features can be explained through combinatorial mechanisms under simple chemical control or if higher-level control mechanisms are required. One hypothesis proposes a way to replicate specific sequences of interunit linkages through a direct template polymerization mechanism. According to this model, an antiparallel double-stranded lignin template, maintained in a dynamic state at the leading edge of each lignifying domain, determines the configuration of the daughter chain being assembled on the proximal strand's exposed face. Furthermore, replication fidelity could be controlled by strong nonbonded orbital interactions between matching pairs of aromatic rings in the parent and the growing daughter chains. The overall process seems to be consistent with the lack of both crystallinity and optical activity in macromolecular lignin domains.

Figure A. Contemporary View of Lignin Substructures.
Theory proposed by G. Brunow and coworkers in 1998 (reproduced with permission).

Finally, required sequence information may be encoded in polypeptide chains that embody arrays of adjacent lignol-binding sites analogous to those found in dirigent positioning proteins.[2]

CITED REFERENCES

[1] J. Ralph et al. (2004). "Lignins: Natural Polymers from Oxidative Coupling of 4-Hydroxyphenylpropanoids," *Phytochemistry Rev.* **3**, 29–60.
[2] S. Sarkanen. (1998). "Template Polymerization in *Lignin Biosynthesis,*" pp. 194–208 in *Lignin and Lignan Biosynthesis* 697, ed. N. G. Lewis and S. Sarkanen, American Chemical Society, Washington, D.C.

Technical Milestones

Within 5 years

- Initiate metagenomic surveys in environments known to break down biomass naturally, and identify and molecularly characterize ligninases and hemicellulases, including accessory enzymes.
- Heterologously express enzymes in suitable protein production hosts.
- Determine catalytic properties of enzymes.
- Use overproduced and commercial enzymes to digest native hemicelluloses, and identify intermediate products in hemicellulose decomposition.
- Perform molecular evolution of enzymes, and select for improved kinetic efficiency and compatibility with other enzymes and with desired industrial applications. This is expected to result in the identification of factors limiting enzymatic conversion and the development of enzyme cocktails to overcome them efficiently.

Within 10 years

- Extend metagenomic surveys to discover novel ligninases and hemicellulases.
- Concurrently improve enzymes in the cell walls of energy crops in parallel with knowledge about substrate-imposed limitations.
- Develop molecular models for enzyme-substrate structure-function relationships.
- Improve catalytic efficiency and stability of enzymes that degrade cell walls using protein engineering.

Constraints imposed by the fairly limited set of industrial host organisms on expression of native or heterologous GH proteins must be understood (for example, the role and modification of various glycoforms). Greater understanding in these areas would allow more efficient improvements in enzyme performance in industrialized hosts and resultant production economics. Understanding biomass recalcitrance will enable enzyme preparations to be engineered for superior performance using proteomics and the systems biology approach. These challenges will be more pronounced when dealing with "foreign" proteins (such as ligninases) and multienzyme complexes (cellulosomes) in heterologous hosts.

Within 15 years

- Perform rational engineering of enzymes to specific requirements based on new understanding of enzymes and substrate cellulose-hemicelluloselignin complexes. The ability to design specific enzymes to attack specific substrates is expected by this time.
- Greatly increase understanding of fundamental protein-secretion limitations in hyperproducing strains. Similar challenges must be met to enable consolidated bioprocessing microbes.

- Similarly, the catalytic activity and thermal stability of ligninases must be enhanced through directed evolution strategies, which require identification of suitable model substrates for screening mutant enzymes in variant libraries.

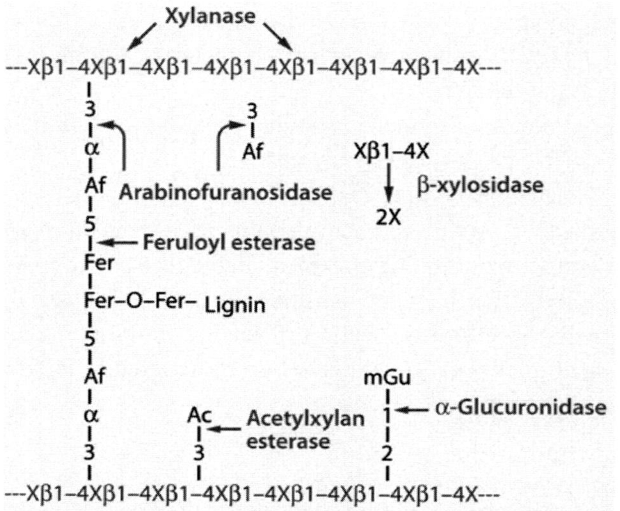

```
              Xylanase
         ↙            ↘
---Xβ1-4Xβ1-4Xβ1-4Xβ1-4Xβ1-4Xβ1-4Xβ1-4X---
    |                    |
    3                    3
    |                    |
    α                    Af         Xβ1-4X
    |                               ↓
    Af  Arabinofuranosidase       β-xylosidase
    |                               2X
    5
    | ← Feruloyl esterase
    Fer
    |
    Fer-O-Fer- Lignin
    |
    5
    |
    Af
    |                              mGu
    α         Ac    Acetylxylan     | ← α-Glucuronidase
    |          |    esterase        |
    3          3                    2
    |          |                    |
---Xβ1-4Xβ1-4Xβ1-4Xβ1-4Xβ1-4Xβ1-4X---
```

Source: Molecular structure adapted from L. B. Selinger, C. W. Forsberg, and K. J. Cheng, "The Rumen: A Unique Source of Enzymes for Enhancing Livestock Production," Anaerobe 2(5), 263–84 (1996).

Figure 2. Complex Mixture of Enzymes for Degrading Hemicelluloses.
The example depicted is cross-linked glucurono arabinoxylan. The complex composition and structure of hemicellulose require multiple enzymes to break down the polymer into sugar monomers—primarily xylose, but other pentose and hexose sugars also are present in hemicelluloses. A variety of debranching enzymes (red) act on diverse side chains hanging off the xylan backbone (blue). These debranching enzymes include arabinofuranosidase, feruloyl esterase, acetylxylan esterase, and alpha-glucuronidase (see table below). As the side chains are released, the xylan backbone is exposed and made more accessible to cleavage by xylanase. Beta-xylosidase cleaves xylobiose into two xylose monomers; this enzyme also can release xylose from the end of the xylan backbone or a xylo-oligosaccharide. The table shows that some of these enzymes are multifunctional, with catalytic domains belonging to different enzyme families. Their great diversity and that of other enzymes involved in hemicellulose degradation present a remarkably complicated enzymatic system whose more-thorough analysis may yield greater understanding of hemicellulosic degradation.

Glycoside Hydrolase (GH) and Carbohydrate Esterase (CE) Enzyme Families for Degrading Hemicelluloses

Enzyme	Enzyme Families
Endoxylanase	GH5, 8, 10, 11, 43
Beta-xylosidase	GH3, 39, 43, 52, 54
Alpha-L-arabinofuranosidase	GH3, 43, 51, 54, 62
Alpha-glucuronidase	GH4, 67
Alpha-galactosidase	GH4, 36
Acetylxylan esterase	CE1, 2, 3, 4, 5, 6, 7
Feruloyl esterase	CE1

Crosscutting Tools, Technologies, and Science

Computational science, advanced imaging, and high-throughput genomics tools must be incorporated to attain the objectives described here. Computational science is important in coordinating process engineering and in exploiting the products of protein engineering. Structural imaging of enzymes, substrates, and enzyme-substrate complexes will be needed to understand enzyme function at the molecular level; a library of model substrates for these kinetic and mechanistic studies is required.

A better understanding of the structure and composition of specific xylans and lignins would greatly improve the ability to design optimal enzyme mixes. Methods to monitor the process of polysaccharide hydrolysis (e.g., NMR and mass spectroscopy) would be beneficial. Transgenic plants and trees with reduced lignin contents or modified lignin configurations could exhibit differences in lignin-depolymerization rate; understanding enzymatic mechanisms would be crucial in manipulating conditions to the best effect.

Advanced imaging, spectroscopic, and enzymatic methods of characterizing cell walls before and after pretreatment also require new resources for producing and disseminating chemical standards. These new standards will be representative of the range of plant cell-wall chemistry and will permit improved cross-comparison research, not only on polysaccharide disassembly but also on subsequent sugar metabolism.

UNDERSTANDING THE MOLECULAR MACHINERY UNDERPINNING CELLULOSE SACCHARIFLCATION: CELLULASES AND CELLULOSOMES

With new biology tools, we have the opportunity to understand enzyme structure-function relationships that govern the critical processes of bioethanol production. Given the importance of alternative renewable energy sources, the prospect of engineering improved cellulases is an exciting and universally appealing concept. Cellulosomes, large molecular machines that integrate numerous enzymes and functions to break down biomass, are key to reducing the enzyme loading required for processing. Understanding cellulosomes and learning how to manipulate and modify them for greater efficiency will be important in consolidated processing. Research first will allow understanding and then improvement in the performance of both free and complexed (cellulosomal) cellulases on biofeedstocks. The rate-limiting step in hydrolysis is not catalytic cleavage but disruption of a single substrate chain from its native matrix, thereby rendering it accessible to the catalytically active cellulase site. Thus, we must be able to analyze and understand the processes and interactions that facilitate this disruption of insoluble cellulose. To approach this problem in a systematic and rational fashion, detailed understanding of the structure and energetics of both the micro-crystalline and noncrystalline portions of cellulose fibrils is first necessary.

- How do soluble enzymes act on an insoluble crystalline substrate? Hydrolysis of crystalline cellulose is the rate-limiting step in biomass conversion to ethanol because aqueous enzyme solutions have difficulty acting on this insoluble, highly ordered structure. Cellulose molecules in their crystalline form are packed so tightly

that enzymes and even small molecules such as water are unable to permeate the structure.
- How do different biomass-degrading enzymes work together as a synergistic system? Cellulases and hemicellulases are secreted from cells as free enzymes or as extracellular cellulosomes. The collective activity of enzyme systems is believed to be much more efficient than the individual activity of any isolated enzyme; therefore, to truly understand how enzymes function, they must be studied as systems rather than individually or a few at a time. In addition, systems eventually must be analyzed under laboratory conditions more representative of real-world environments. For example, laboratories often use purified cellulose as the substrate for enzyme analysis rather than more heterogeneous, natural lignocellulosic materials; this can provide erroneous conclusions about natural enzyme activity. New analytical methods that are spatially and chemically sensitive will allow realistic understanding of the mechanisms of biomass degradation.

Research Goals

Discovering and Improving Free Cellulases

New generations of engineered cellulases will provide enhanced performance (activity) requiring lower protein loadings, so process cost will be reduced. Cellulase cost currently is estimated at 10 to 25 cents per gallon of ethanol produced. The new target will be similar to starch hydrolysis, or 1 to 2 cents per gallon of ethanol. For example, loadings of about 25- mg fungal cellulase proteins are required to hydrolyze about one gram of cellulose in pretreated biomass. Improved cellulases would convert an equivalent amount of cellulose to sugars using around one-tenth the enzyme loading. More efficient cellulase biomass digestion also may permit reduced-severity thermochemical pretreatment.

Acquiring new scientific insights into cellulose structure and the function of cellulase cocktails is a critical objective (see sidebar, New Cellulase Enzymes Dramatically Reduce Costs of Plant Biomass Breakdown, p. 100). The initial research phase involves an exhaustive search for examples of important families of free cellulases (those secreted into the extra-cellular mixture) using high-throughput genomic and enzyme-production and -characterization methods. Cellulose and cell-wall interaction with degrading enzymes must be characterized. Biochemical analysis of family members will reveal much about the natural diversity of solutions to cellulase action. Combined mutational analysis and computational modeling then will be used to define structure-function relationships of these enzymes and newly engineered mutants. With this kinetic and thermodynamic understanding as a guide, enzyme-engineering approaches will be used to test novel hypotheses for improving cellulase performance on the cell walls of plants considered suitable biofeedstocks.

Understanding and Utilizing Cellulosomes

The cellulosome, a unique type of molecular machine, can efficiently solubilize native and pretreated lignocellulose substrates. The cellulosome serves as a more efficient way of enzyme loading and presentation to the substrate. In this case, native cellulosomes attached to the microbe contain the full repertoire of plant cell-wall polysaccharide-degrading enzymes, and a single carbohydrate-binding module (CBM) targets the entire complement of enzymes to the substrate surface (see sidebar, The Cellulosome, p. 102).

- Artificial designer cellulosomes that exhibit a precise architecture and composition will help reveal various principles of cellulosome construction and action. In this context,
    Structure and biochemical-biophysical aspects of assorted cellulosomal modules (i.e., cohesins, dockerins, CBMs, and catalytic domains) from different subunits and species will be examined.
    Recombinant primary and adaptor chimeric scaffoldins will be constructed, designed to bear divergent cohesins for subsequent incorporation of native or hybrid (dockerin-tagged) cellulosomal enzymes.
    Through this approach, improved high-performance (noncellulosomal) cellulases can be incorporated into a cellulosome to take advantage of enhanced synergistic properties inherent in the cellulosome complex.

Genes encoding for the complement of designer cellulosomal components also can be cloned into a suitable host cell system for heterologous production, assembly, and secretion of active designer cellulosomes of desired composition and architecture.

NEW CELLULASE ENZYMES DRAMATICALLY REDUCE COSTS OF PLANT BIOMASS BREAKDOWN: R&D 100 AWARD

Further Advances Needed to Improve Efficiency and Economics

Cellulase enzymes are used to break down the cellulose of plant cell walls into simple sugars that can be transformed (fermented) by microbes to fuels, primarily ethanol, as well as to chemicals, plastics, fibers, detergents, pharmaceuticals, and many other products.

Like starch and sugar, cellulose is a carbohydrate (compound of carbon, hydrogen, and oxygen) made up of simple sugars (glucose) linked together in long chains called polysaccharides. These polymers form the structural portion of plant cell walls, and unraveling them is the key to economical ethanol fermentation. Technical barriers to large-scale use of cellulose technology include the low specific enzyme activity and high enzyme-production costs, as well as a general lack of understanding about enzyme biochemistry and mechanistic fundamentals.

In 2004, the DOE National Renewable Energy Laboratory (NREL), working with two of the largest industrial enzyme producers (Genencor International and Novozymes Biotech), achieved a dramatic reduction in cellulase enzyme costs. Cellulases belong to a group of enzymes known as glycosyl hydrolases, which break (hydrolyze) bonds linking a carbohydrate to another molecule. The new technology involves a cocktail of three types of cellulases—endoglucanases, exoglucanases, and beta-glucosidases. These enzymes work together to attack cellulose chains, pulling them away from the crystalline structure and breaking off cellobiose molecules (two linked glucose residues), splitting them into individual glucose molecules, and making them available for further processing. This breakthrough work resulted in 20- to 30-fold cost reduction and earned NREL and collaborators an R&D 100 Award (press release: www.nrel.gov/awards/2004hrvtd.html).

Cellulase image from M. Himmel et al., "Cellulase Animation," run time 11 min., National Renewable Energy Laboratory (2000).

Further cost reductions are required, however, to support an economical and robust cellulose biorefinery industry. For example, costs of amylase enzymes for converting corn grain starch to ethanol are about 1 to 2 cents per gallon of ethanol produced, but the most optimistic cost estimates for cellulase preparations now are about tenfold higher than that. Routes to improving enzyme efficiencies include the development of enzymes with more heat tolerance and higher specific activities, better matching of enzymes and plant cell-wall polymers, and development of high-solid enzymatic hydrolysis to lower capital costs. A comprehensive understanding of the structure and function of these enzymatic protein machines, how their production and activity are controlled, and changes they promote on plant cell-wall surfaces will be critical for success.

(Also see sidebar, Image Analysis of Bioenergy Plant Cell Surfaces at the OBP Biomass Surface Characterization Lab, p. 40.)

Several key scientific questions and issues are especially important for reaching the proposed goals.

- We do not understand enough about substrate microcrystalline cel lulose's structure as it exists in the plant cell wall. Uncertainties about the crystal lattice structure in the microfibril and the crystal faces of cellulose targeted by enzyme diversity in the biosphere are examples of our lack of insight. Understanding the substrates' molecular architec ture would help in using rational design to produce improved hydro lytic molecular machines. With such knowledge it might be possible, for example, to select specific mutations of cellulose-binding domains that tend to disrupt fibril packing or to use molecular dynamic simulations to find environmental conditions (temperature, pressure, and molecular agents) that affect such disruption before the cellulase attacks the substrate.

- The kinetic and thermodynamic mechanism of processive cellulases or exoglucanases (GH family 7) is not known. These enzymes, protein machines, conduct most cellulose hydrolysis, with endoglucanases playing a lesser role. To understand this mode of catalysis at the molecular level, a mathematical model of the functioning of GH family 7 cellulase must be developed and tested. Finally, the range of permissible modifications of these specific protein folds is not known, nor are constraints imposed by the fairly limited set of industrial host organisms on expression of wild-type members of GH families or newly engineered mutants.
- The major objective of the designer-cellulosome concept is to reconstruct improved cellulosomes by linking essential and most efficient enzymes to desired substrates. Numerous scientific issues and opportunities are still "at large."

    Our capacity to control the final designer-cellulosome's composition and architectural arrangement will enable us to pose new hypotheses regarding enhanced cellulosome performance in degradation of plant cell-wall material.

    The best set of cellulosomal cellulases most appropriate (synergistic) for use on lignocellulosic substrates is still unknown.

    Although hybrid enzymes derived from noncellulosomal species can be incorporated into designer cellulosomes, whether their observed synergistic activity is comparable to that of native cellulosomal enzymes is still unknown.

    Many other scientific questions have to be addressed, including the optimal stoichiometry and arrangement of components and whether the combined action of cellulosomal and noncellulosomal systems results in an additional improvement in performance. This research offers a unique opportunity to explore the expanding universe of molecular machines that can be constructed by the designer-cellulosome approach.

Technical Milestones

Within 5 years

- Near-term accomplishments in cellulase biochemistry should focus on improving the enzymes, primarily from fungi and actinomycetes, used in industrial preparations marketed for the biorefinery. This accomplishment would require searching for new enzymes from new sources (including metagenomic databases) and using directed evolution to probe mutational space for possible improvements over wild-type examples. The outcome would be better cocktails based on wild-type enzymes and enzymes improved by noninformational methods that could be used immediately by industry.
- Construction of designer cellulosomes will concentrate on several fronts, including incorporation of currently available enzymes; development of novel cellulosomal components—scaffoldins, cohesins, dockerins, CBMs, and linker segments; improvement of hybrid enzymes and assessment of their enhanced synergistic action within designer cellulosomes; and analysis of the combined action of designer cellulosomes with highly active noncellulosomal enzymes, including ligninases and hemicellulases.

Within 10 years

- The mid-term scope permits a more systematic approach to cellulase and cellulosome biochemistry. Such understanding will require combining classical biochemistry with computational science.
- A particularly important goal is identifying thermodynamic limitations of improvement, which would impact protein engineering.
- Initial studies to convert suitable host-cell microorganisms into cellulosome-producing strains will be pursued. The desired result will be improved biomass degradation and increased understanding of the structure-function relationship of cellulosome components.
- To achieve heterologous production, assembly, and secretion of active designer cellulosomes, suitable host cells that can accommodate the genes for such large proteins must be identified and inherent cloning and expression barriers overcome.

Within 15 years

- HTP approaches and crystallographic and microscopic techniques will be used to elucidate the three-dimensional structure of cellulosomal components, the intact native and designer cellulosomes, and their action with pure (cellulose) and native (plant cell-wall) substrates.
- Understanding mechanistic principles of the entire accessible range of glycoside hydrolases and their function in the biosphere will be evaluated. The diversity of cellulase and cellulosome functional schemes will be modeled and optimized for specific biomass substrates (feedstocks). This improved understanding of cellulase action will provide new saccharification paradigms for the biorefinery.

Crosscutting Tools, Technologies, and Science

Computational science, advanced imaging, and HTP genomic tools must be incorporated to attain the objectives described here.

- Computational science is needed for understanding cellulase function at a holistic level, specifically, new programs and codes to handle models of ultralarge biological molecules and systems of more than 1 million atoms.
- New imaging methods are critical for elucidating cellulase action on the plant cell wall.
- Finally, tools to acquire, archive, and interpret new cellulase structures (both native and modified) are needed to better understand their natural structural and functional diversity.

We must develop new science and technical methodologies to explore protein-machine function at the molecular and atomic levels to attain these objectives. Solution physics and thermodynamics ruling protein-domain biological function at the nanometer scale are poorly understood. In this respect, the new field of nanoscience may provide important insights. For example, describing the mechanism of a processive cellulase at such level of detail will require more knowledge about the cellulose surface, water-protein-glycan dynamics, and biomolecular mechanics than is possible today.

The Role of GTL and OBP Facilities and Capabilities

Protein Production

Protein production capabilities will be critical for supplying samples of wild-type and recombinant proteins, including (1) new enzymes from new sources, (2) modified enzymes from directed evolution, (3) modified enzymes from site-directed mutation, and (4) large quantities of enzymes suitable for pilot testing. This resource will be critical in identifying the most effective new hemicellulases and ligninases without first having to solve high-level expression challenges for all candidates.

High-throughput (HTP) biochemical and biophysical assays will be developed and carried out, and functional annotation will enrich the first-pass sequence-based annotations. This problem is more difficult and complex for GH, in which a single enzyme can have several comparable activities on different polysaccharide substrates. Thus, a microchip HTP assay facility is needed.

Molecular Machine

These capabilities can provide analytical and computational science to study cellulase protein-machine function and advanced techniques for isolating and reconstituting complex interactions within native or improved hemicellulase and ligninase systems. Synergistic interactors including chaperonins, excretion paths, and cofactors will be isolated and identified in novel cellulases, hemicellulases, and ligninases.

Proteomics

Proteomic capabilities can be used to document how secretomes of selected white rot fungi and other organisms such as microorganisms in the termite hindgut respond to changing culture conditions and optimize biodegradation rates of lignocellulosic components. Correlations between ligninase activity and concentrations of enzymes responsible for cosubstrate production will be of particular interest and will assist researchers in designing more effective experimental plans for improving the energetics and carbon-allocation efficiency of cell-wall degradation. This also will be important in monitoring and controlling the effects of substrate inducers and responses to heterologous expression, especially of the most "foreign" components. This resource can support studies aimed at improving production of ligninase and hemicellulase proteins from both near-term hosts such as T. reesei and longer-term systems including bacteria and yeast.

Cellular System

Cellular system capabilities will visualize cellulases, ligninases, cellulosomes, and other molecular species directly on substrates in real time under varying conditions and from varying species. Models will be developed to optimize hydrolysis under different conditions.

DOE Joint Genome Institute

DOE JGI could embark on sequencing new white rot fungal genomes, including those that do not encode lignin and manganese-dependent peroxidases. This work may lead to the discovery of more-active and less-labile ligninase variants. After that, more biological diversity from such bacterial populations as decaying biomass, termite hindgut, ruminants, and the rhizosphere could be examined.

Metagenomics

Metagenomic approaches will identify mechanisms and agents of effective hemicellulose and lignin conversion in communal or unculturable populations. They also can play a key role in sequencing new genomes thought to harbor cellulolytic components acquired by lateral gene transfer and in studying natural adaptations that allow (initially) heterologous expression.

THE CELLULOSOME: THE "SWISS ARMY KNIFE" OF MOLECULAR MACHINES

The cellulosome is an extracellular supramolecular machine synthesized by some anaerobic microorganisms capable of degrading crystalline cellulose and other plant cell-wall polysaccharides. Each cellulosome contains many different complementary types of carbohydrate-active enzymes, including cellulases, hemicellulases, and carbohydrate esterases that are held together by a scaffoldin protein to form a single multienzyme complex (see Figure. A. Schematic of a Cellulosome, below). The cellulosome enhances cell-wall degradation by bringing several different enzymes into close proximity so they can work together to exploit enzyme-accessible regions of cellulose. The various product intermediates from one enzymatic subunit can be transferred readily to other enzymatic subunits for further hydrolysis (breakdown) of the cellulose. The cellulosome also promotes cell adhesion to the insoluble cellulose substrate, thus providing individual microbial cells with a direct competitive advantage in using soluble sugars produced in the hydrolysis. Cellulosomes need not be associated with cells for activity, and they function under both aerobic and anaerobic conditions.

Each enzymatic subunit contains a definitive catalytic module and a dockerin domain that binds tightly to a scaffoldin cohesin. Thus, cohesin-dockerin interaction governs assembly of the complex, while cellulosome interaction with cellulose is mediated by scaffoldin-borne cellulose-binding molecules (CBM). Some scaffoldins also bear a divergent type of dockerin that interacts with a matching cohesin on an anchoring protein, thereby mediating cellulosome attachment to the cell surface.

The LEGO-like arrangement of cellulosomal modules offers an excellent opportunity to engineer new multi- enzyme complexes for desired purposes. The various cohesins (c), dockerins (d), and catalytic modules (A, B, C) are functionally independent and can be tethered together in any combination via recombinant genetics. The resulting complex can be applied either in the test tube (bioreactors) or in a cellular setting (fermentors).

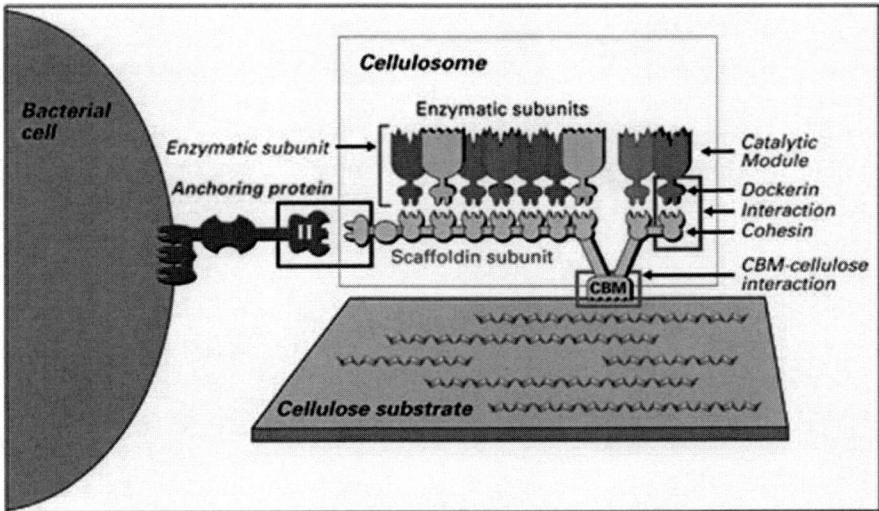

Figure A. Schematic of a Cellulosome.
The scaffoldin subunit, to which the cellulosomal enzymatic subunits are bound, is made up of different functional modules that dictate its architecture and various activities. These include a single CBM and nine very similar repeating domains called cohesins, which bind the enzymatic subunits.

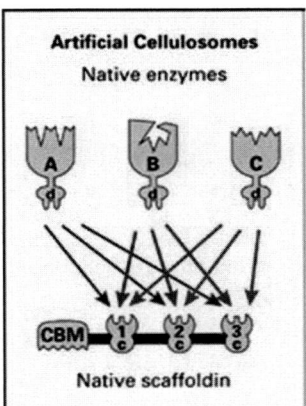

Figure B1. Random Incorporation.

Native scaffoldin-borne cohesins in cellulosome-producing bacteria generally recognize all enzyme-bearing dockerins nonspecifically. Thus, reconstruction of a cellulosome from its component parts would result in random incorporation of different enzymes into cellulosome complexes, yielding a heterogeneous population of artificial cellulosomes (Figure. B1. Random Incorporation, below).

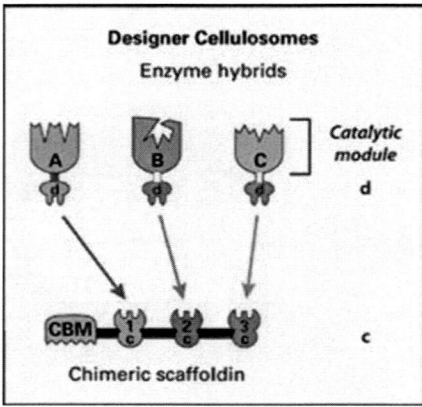

Figure B2. Controlled Incorporation.

To control the incorporation of desired enzymes into a precise position, distinctive cohesin-dockerin pairs must be used (Figure. B2. Controlled Incorporation, below.). Dockerin domains that bind a specific cohesin must be fused to the different catalytic modules, and a chimeric scaffoldin is engineered to contain complementary cohesins and a single CBM for substrate targeting. Subsequent self-assembly of the mature "designer" cellulosome complex can then occur selectively in vitro, resulting in a homogeneous population of cellulosome complexes.
Cellulosome References: p. 117

HARVESTING THE BIOCHEMICAL POTENTIAL OF MICROORGANISMS THROUGH METAGENOMICS

The conceptual foundation for metagenomics is the realization that more than 99% of microbes in most habitats have not yet been cultured. By providing deep insight into biological capabilities and function without the need for culturing, metagenomics effectively expands our scientific capabilities and understanding beyond the small percentage of bacteria that currently can be cultivated in the laboratory. It further permits a more holistic and mechanistic analysis of microbial communities. Natural systems long ago learned how to attack lignocellulose and use the resultant sugars and other chemicals for their own purposes. The metagenomes of complex natural communities provide a fertile resource for data mining to search for new examples of relevant enzymes. The ability to construct and evaluate an engineered metagenome (from existing genome sequences) now enables creation of systems with superior capabilities.

This research has two objectives: (1) employ genetic and biochemical studies to characterize novel lignocellulose-degrading systems and (2) use metagenomics to augment these efforts by characterizing and recovering the genetic potential (including that from uncultured microbes) resident in microbial communities capable of rapid and extensive biomass degradation. Outcomes of these functional and comparative studies will include a repertoire of new enzymes and proteins available for engineering approaches (e.g., designer cellulosomes or free cellulase systems). In that context, knowledge and bioresources arising from this research are well integrated within the overall goal of improving plant cell-wall deconstruction and conversion of lignocellulosic products to bioethanol.

Research Goals

To elucidate the repertoire of gene products necessary to effect more rapid and extensive hydrolysis (and solubilization) of lignocellulosic materials, systematic advances in our understanding of microbe-mediated plant cell-wall deconstruction must be achieved. First, many sequenced microbial genomes encode more glycoside hydrolases for plant cell-wall deconstruction than have been identified through historical biochemical and genetic studies. Whether or not these multigene families are an evolutionary adaptation to provide more "total" activity diversity, to compensate for subtleties in substrate conformation and composition, or a combination of both is still unknown. In addition, degradative synergism in microbial communities often resides in the concerted actions of enzymes from multiple organisms, but mechanistic details and understanding of this concept are undeveloped.

Second, aerobic and anaerobic bacteria with superior degradation capabilities recently have been found to lack canonical processive cellulases required in all other well-described systems for substrate solubilization. The genetics and biochemistry underpinning their lignocellulose degradation, which could represent a new mechanism, must be explored.

Technical Milestones

GTL will accelerate the development of optimal cellulase systems by providing resources for screening thousands of natural and modified enzyme variants, enabling the HTP production and functional analysis of these enzymes, elucidating regulatory controls and essential molecular interactions, and generating models for analyzing the structure and activity of natural and engineered enzyme systems.

Within 5 years

- Undertake microbiome projects for functional and comparative purposes. The JGI Community Sequencing Program, supported by DOE, is now characterizing the microbiome of the termite hindgut.
- Initiate mechanistic characterization (at DNA, protein, and organismal levels) of no fewer than three natural microbial communities displaying rapid and extensive lignocellulose degradation. Each should be different in terms of prevailing physicochemical conditions (e.g., temperature, pH, and salt concentration) or dominant feedstock. For example:
  Soils representative of dominant residues in the DOE-USDA Billion-Ton study. Corn stover, rice, switchgrass, and forest residues would be appropriate.
  Decaying biomass (see the sidebar, Examples of Metagenomic Analysis: Understanding the Dynamics of Microbial Colonization of Decaying Biomass, p. 108).

- Determine the mechanisms of lignocellulose solubilization by such known organisms as *F. succinogenes* and *C. hutchinsonii* to help guide metagenomic analyses, using a combination of biochemical and genomic approaches (see sidebar, Sequencing a Soil-Cellulose Degrader, this page).
- Develop heterologous expression systems for candidate enzymes and regulatory proteins, including necessary cofactors and post-translational modifications.
- Begin HTP production, characterization, and intercomparison of enzymatic systems discovered in metagenomic analyses. Produce appropriate molecular tags to support experimentation in natural and controlled systems—including imaging, protein isolation, and detection of interactions.

Within 10 years

- Complete comprehensive analyses of enzymatic systems to isolate different families and high performers.
- Rationally modify, express, and characterize native enzymes to understand design principles and optimize properties.
- Build computational models of biomass-decay microbial communities, and test against experimental data.

Within 15 years

- Deploy innovative processes based on these discoveries and bioresources to reduce enzyme costs and loading by tenfold. Provide new options that reduce or eliminate the need for nonbiological feedstock pretreatments.

The Role of GTL and OBP Facilities and Capabilities

In brief, the facilities will support and enable the attainment of these objectives. For instance, JGI already has contributed a foundational set of genome sequences for specialist cellulolytic microbes; this data set is augmented by genomes sequenced at The Institute for Genomic Research (called TIGR). Other resources can readily support hypothesis-driven research at the RNA, protein, organismal, and community levels (see sidebar, Example of Metagenomic Analysis, this page). This research will validate and produce the desired combination of proteins and other biologics necessary to make lignocellulose conversion comparable to starch as a feedstock in economic and process criteria.

SEQUENCING A SOIL-CELLULOSE DEGRADER

The DOE Joint Genome Institute is sequencing *Cytophaga hutchinsonii,* an aerobic Gram-negative bacterium commonly found in soil that rapidly digests crystalline cellulose. Molecular analysis of cellulose degradation by *C. hutchinsonii* is now feasible, since techniques for genetically manipulating the organism recently have been developed.

This microbe exhibits the ability to move rapidly over surfaces by a process known as gliding motility, which is thought to be important in allowing *C. hutchinsonii* to colonize its insoluble growth substrate. The mechanism of gliding motility is not known, but flagella are not involved. Analysis of the *C. hutchinsonii* genome sequence will facilitate studies of cellulose degradation and also will reveal more about bacterial gliding motility, which has remained an unsolved biological mystery for over 100 years (http://genome.jgi-psf.org/finished_microbes/cythu/cythu.home.html).

EXAMPLE OF METAGENOMIC ANALYSIS: UNDERSTANDING THE DYNAMICS OF MICROBIAL COLONIZATION OF DECAYING BIOMASS

As primary decomposers, microbial communities have evolved both as competitors and collaborators in biomass deconstruction. The ultimate aim of this research is to achieve bioprocesses for all steps in converting biomass to ethanol. A critical need is to replace plant biomass thermochemical pretreatment, which is now necessary to convert recalcitrant structural lignoellulose to a form in which cellulose is more accessible and amenable to hydrolytic enzyme action. By analyzing natural communities that colonize decaying biomass, we can ascertain natural mechanisms that can be used to supplant thermochemical treatments. The goal is to better understand the complex microbial communities responsible for lignocellulosic biomass deconstruction, harvest key biochemical decay mechanisms, and develop predictive modeling and control of these complex natural processes.

Metagenomics will be used to determine microbial-community composition and genetic diversity. Comparative genomic tools will initially determine community functionality and identify genes of unknown function. Coupled spatial and temporal measurements will reveal principles of community formation and dynamics, processes such as signaling, and other microbial interactions. Functional annotation of unknown proteins through protein production and characterization and other experimentation will augment initial gene-function assignment.

HTP methods from facility and other capabilities will be used for measuring enzyme activities:

- Determine cellular and biochemical functions of genes discovered in uncultured community members.
- Characterize the temporal composition and functional capability of microbial communities—Who is there, and what metabolic processes are being carried out?
- Characterize expression patterns of cellulolytic enzymes and related processes and pathways using transcriptomics, proteomics, and metabolomics.

Computational tools will predict the metabolic, physiologic, and behavioral characteristics of microbial communities from community DNA sequence data, and supporting measurements will be developed. The tools will allow design and engineering of microbial systems that ultimately can perform all steps in biomass processing to ethanol.

CHARACTERIZING CELL WALLS USING HIGH-THROUGHPUT METHODS

Understanding the chemical nature and architecture of cell walls at the nano-, micro-, macrochemical, and physical scales, as well as their behavior in pretreatment and fermentation, is essential in taking a systems-level approach to modifying the plant genome or increasing the biofuel system's efficiency. Knowledge and insight gained from HTP cell-wall characterization will drive several areas of biomass-to-biofuels research. These areas include selecting and modifying plants, matching the microbial conversion process to the substrates present, and minimizing inhibitors during biological conversion to valuable products. Many new instruments and analytical methods will be needed to characterize a range of biomass feedstocks (e.g., grasses, crop residues, short-rotation woods, and early energy crops) needed to achieve the Billion-Ton vision. Ultimately, many different types of feedstocks will be at different stages of harvest and hence will need to be analyzed rapidly. The robustness of these methods and their versatility in addressing the composition of many different materials requires significant further research. Data generated by these methods are used to value feedstocks, measure conversion efficiency, identify regulatory issues, and ultimately establish product pricing and investment risk.

Research must address the disconnect between the throughput of genomic and proteomic analysis and that of biomass chemical and structural characterization. DNA sequencing and metabolic profiling are routinely done at the microliter scale, where 1000 to 2000 samples can be processed and analyzed in one day. Most biomass analysis requires hundreds of milligrams per sample and very few have been automated, so throughput is around 20 samples per week per person. Recent advancements in biomass analysis have demonstrated high throughput using multivariate analysis (MVA) coupled with analytical pyrolysis, FTIR, or near infrared spectroscopy to allow characterization of 200 to 500 samples per day per person. These spectroscopic methods, however, require calibration and validation using slow wet chemical procedures. MVA often is specific to feedstock and process; for example, a method developed for hardwood analysis won't accurately characterize a softwood or grass sample. To support the broad goals of this roadmap, many new analytical methods will need to be developed.

Research will focus on two areas: Improving throughput of traditional calibration methods to around 500 samples per week and using thse data to calibrate "next-generation" capabilities of analyzing 1000 or more samples per day. Strategies for improving traditional methods will require the adaptation of analytical instruments for biomass applications and development of new instruments specifically for this work. For example, multiplexing robotic sample preparation and multichannel capillary electrophoresis in carbohydrate analysis potentially could offer a tenfold increase in sample throughput, but such instruments and methods are not validated for biomass samples. New methods and instruments will generate data on molecular- and genetic-level relationships needed to (1) improve feedstocks by increasing biomass production and making feedstocks more amenable to the next generation of biomass-conversion technologies, and (2) develop data sets that enable the design of microbial cell conversion strategies that will produce desired products in high yield and purity. A special category of analytical methods with high sensitivity also will be needed to interpret high-resolution images of feedstocks and process substrates.

The complexity of the biomass matrix presents significant analytical challenges not faced in conventional genomics work. Chemical characterization methods will have to assess the relative weight percent of about 14 constituents in each feedstock. Structural constituents of potential interest are protein, lignin, ash, glucan (cellulose), xylan, mannan, arabinan, galactan, glucuronic acid, ferulic acid, acetyl groups, starch, and galacturonic acids (pectin). Some nonstructural materials, which will be specific to each feedstock, also may be of interest in saccharification and fermentation studies. These constituents include inorganic salts, sucrose, tars, waxes, gums, lignans, and others. Analytical methods also will be needed to track the constituents and their reaction products through pretreatment, saccharification, and fermentation. Structural characterization methods must quantify functional groups of interest (e.g., carboxylic acids, ketones, aldehydes, esters, and methoxyl and free-phenolic hydroxyl groups). With the requirement for so many measurements, a portfolio of complementary methods will be needed to close mass balance of feedstock and process components across multiple processing steps.

This research program addresses a critical component in the feedstock-sugars interface by providing analytical instruments and methods with the precision, accuracy, and throughput required to optimize biomass selection and development, biomass pretreatment, and conversion processes leading to economical ethanol production in biorefineries. Biomass-relevant analytical tools of this type do not now exist for these applications.

Tracking individual biomass components from feedstock to products requires high-accuracy data obtained in an integrated and consistent fashion. A research program focused on analytical methods that coordinate data through all stages will improve process integration through consistent data on feedstocks, substrates, process inhibitors, and products. Integrated data will furnish a systems-level understanding of process streams and enable the desired correlations between reactivity and genome expression.

To enable future development of high-energy biomass feedstocks, method portfolios need to be standardized; validated in a consensus environment; and published in a forum available to plant breeders, field scientists, process engineers, enzyme scientists, and fermentation scientists. Technologies will provide data for identifying relationships among plant cell-wall components and interactions among biomass-derived substrates, microbes, and enzymes. Although progress has been made in chemical procedures for biomass analyses, existing portfolios of methods are labor intensive, time consuming, expensive, and generally not amenable to the HTP needs of breeders, agronomists, feedstock processors, and systems approaches to fermentation. To bypass this limitation, HTP techniques need to measure the amounts and structure of cell-wall components in processes. Correlations among process data and genomic and proteomic data will enable identification of genes most important for improved biomass conversion.

For example, more than 225,000 independent T-DNA insertion lines of *Arabidopsis* have been created that represent almost the entire genome space. HTP analysis must be able to screen these large sample sets, as well as the much larger number of samples produced for other species with more complicated genomes, to determine the role each gene plays in different cell-wall chemistry phenotypes. HTP methodologies also will enable screening of plant, enzyme, and microbe consortia to allow for deeper understanding of different plant and microbe genome interactions. High-sensitivity methods will integrate with imaging tools to enable spatial compositional and functional group determinations in individual plant cells. These studies require the analysis of hundreds if not thousands of samples and would not be possible without the cost and time saving provided by rapid biomass-analysis methods.

Technical Milestones

Within 5 years

For near-term deployment, portfolios of precise and accurate analytical methods for a wide range of structural and nonstructural constituents and biomass function groups will be developed and validated in a consensus mode and published for general use. Specific techniques and methods will be tailored to provide comprehensive characterization of biomass model substrates (feedstocks). Working closely with commercial suppliers, investigators will create instruments and techniques specific to the needs of biomass-conversion genomics. Data obtained through customized analytical procedures will be used to calibrate and validate rapid MVA methodologies for HTP screening of feedstocks and biomass-derived research samples. This screening will help to select samples of interest from large populations representing species whose genome sequence has been determined. Selection will be based on identifiable differences in phenotypes of major plant cell-wall biopolymers (lignin, cellulose, hemicellulose, and others) important for conversion to fuels and valuable chemicals. Data generated will be used to support the breeding and development of new plant lines. In conjunction with HTP methods, information-technology strategies will capture, analyze, manage, and disseminate resulting data, enabling creation of a national resource for the biofuel and plant biology research community. To the extent possible, samples and data used in calibrating new methods will be archived safely for future applications. Archived samples may be required for bench-marking, new method development, and crossplatform comparisons.

Within 10 years

HTP methodologies will be integrated with other studies to provide a systems-level understanding of how to alter and improve plant cell-wall composition and structure for efficient and economic biofuel production. High-resolution, molecular-specific images enabled by the availability of molecular tags will allow spatial determinations of chemical and structural features across individual cell types and structures.

Within 15 years

Deployment of HTP and high-sensitivity analytical methods and appropriate data-reduction (informatic) techniques will be completed to integrate genomic and proteomic data with cell-wall chemistry information. A systems-level understanding of interactions between plant cell-wall structure and the microbes and enzymes is needed to convert biomass effectively to fuels and chemicals.

The Role of GTL and OBP Facilities and Capabilities

Many underlying principles and HTP techniques (e.g., robotics, informatics, and pattern recognition) for HTP chemical analysis of plants have been or are being developed by the genomic and metabolomic communities. Collaborations with researchers using the resource will require the adaptation and validation of these techniques for biomass-conversion applications. We expect technologies developed for plants (e.g., high-sensitivity spatial detection of biopolymers) to have applications for studying other organisms.

Information obtained from these analytical methodologies will facilitate the determination of plant-gene function in synthesis of plant cell walls and biopolymers, especially genes related to carbohydrate metabolism, polymerization, and modification. Analytical methods tailored to biomass applications will be used to identify protein function related to cell-wall construction. As additional genomes are sequenced, these tools can be used to characterize gene function in other plant lines of interest for biofuel production.

Crosscutting Tools, Technologies, and Science

Many analytical methods that support a systems approach to biomass conversion will require biomass-relevant standards and tags not readily available from commercial sources. Obtaining the needed array of small-molecule standards will require prep-scale isolation of molecules of interest and advanced techniques of carbohydrate and natural product organic synthesis. As we learn more about the plant genome, important intermediate molecules and biomarkers will be revealed. Techniques for isolating and preparing relevant biomass standards will be transferred to the private sector and made available for biomass research and commercial biorefineries.

While much existing robotic instrumentation and data-reduction technologies are available for constructing HTP methodologies, several areas must be addressed to achieve project goals. New techniques for homogenizing and reducing the particle size of biomass samples without degradation will be needed to prepare biomass samples for automated and robotic systems. New modules, equipment, and accessories for existing analytical techniques such as columns for liquid chromatography will have to be adapted specifically for biomass samples. Rapid biomass-analysis methods with large dynamic ranges will detect multiple compounds with high precision, acceptable accuracy, and short acquisition times. Because these methods are often specific to feedstocks and processes, many new ones may be required. Integrated data from different analytical methods into a single comprehensive MVA calibration set will need testing and validation.

Data-reduction and storage methods with predictive platforms for production and processing pipelines are necessary for retrieving and integrating data from breeding or field studies. New "hands-off" bioinformatic methods will combine, analyze, and correlate data (including metabolomic and genomic information) from multiple systems. Multivariate analysis tools will be applied where pertinent to information extraction from complex systems.

Appropriate biomass analysis is a necessary first step in applying systems biology techniques to bioenergy production. In many cases, analytical methods for biomass substrates that demonstrate the required precision, accuracy, and speed are not available. Analysis using existing methods is too slow and expensive for large-scale screening outlined in many basic technology research programs. Before yields of a desired product can be maximized or undesirable side reactions minimized, analytical methods must accurately monitor changes in the chemical constituents. These methods also must be validated in the complex matrix of biomass-conversion streams. In many cases, biomass methods direct the course of research by revealing the presence of important products or the nature of yield-reducing side reactions. Biomass-conversion methods will lead us one step closer to those for accurately monitoring complex degradation reactions in natural systems.

BREAKTHROUGH, HIGH-PAYOFF OPPORTUNITY: SIMPLIFYING THE BIOCONVERSION PROCESS BY UNDERSTANDING CELL-WALL DECONSTRUCTION ENZYMES EXPRESSED IN PLANTS

Scientific Challenges and Opportunities

Complexity is one of the major challenges to economical biomass conversion for large-scale production of liquid transportation biofuels. The process currently requires many steps, including pretreatment, detoxification, solid and liquid separation, cellulase production, cellulose hydrolysis, biomass sugar fermentation, and product recovery supported by such ancillary systems as wastewater treatment and utilities. This complexity results in added operating and capital costs as well as issues of robustness, and simpler procedures would be highly desirable. Although engineering potentially could improve the process incrementally, systems biology approaches have the potential to revolutionize it. Various strategies envisioned to consolidate processing steps will genetically incorporate capabilities within organisms. One potential strategy would include incorporating genes that encode lignocellulose-degradative enzymes into feedstock plants.

- Enzymes could target wall polysaccharides (backbones, side chains, and small substituents that can have big effects on activity of other enzymes), simple phenolic cross-links between polysaccharides such as those that tie together xylans in cereal walls.
- Promoters would be required that are activated from senescence-related or harvest insult–triggered genes or by a signal provided exogenously at a time deemed appropriate.
- Genes encoding enzymes or other disassembly proteins would be required, perhaps those of microbial origin whose discovery would be a major "engineering" goal of the larger project and whose expression would be driven by the promoters above.

Research Goals

While targeting specific objectives using the existing state of technology as a benchmark is important, research must be open to new ideas and concepts. Specific areas that need to be addressed in fundamental and applied areas include the following.

Fundamental Science

- What new learning would enable expression of structural polysaccharide-degrading enzymes in plant tissues? We need new promoters, genes, transformation tools, and knowledge about the sequestration and action of such enzymes when expressed in living plants.
- What is the relative effectiveness of cellulase-enzyme systems acting in cellulose-enzyme-microbe complexes as compared to cellulose-enzyme complexes, and what is the mechanistic basis for such enzyme-microbe synergy?

Applied Science

- Can hemicellulose and cellulose hydrolysis be accomplished without the acidic or alkaline high-temperature and -pressure pretreatment step? This is particularly important for advantageous expression of hemicellulases and cellulase enzymes in plants.
- Can biocatalysts be developed whose performance is not impacted significantly by inhibitory compounds formed as by-products of pretreatment, hydrolysis, or fermentation? If so, they have the potential to eliminate the costly and waste-intensive detoxification step.
- Can such upstream processes as storage (ensilage) or biopretreatment be developed that make cellulose microfibrils more amenable to enzymatic cellulose hydrolysis without the need for relatively harsh thermochemical pretreatment?

REFERENCES

Aden, A., et al. (2002). *Lignocellulosic Biomass to Ethanol Process Design and Economics Utilizing Co-Current Dilute Acid Prehydrolysis and Enzymatic Hydrolysis for Corn Stover*, NREL Report No. TP-510-32438.

Agblevor, F., Chum, H. L. & Johnson, D. K. (1993). "Compositional Analysis of NIST Biomass Standards from the IEA Whole Feedstock Round Robin," pp. 3 95–421 in *Energy from Biomass and Wastes XVI: Proceedings of the 7th Institute of Gas Technology Clean Fuels and Energy from Biomass and Wastes Conference*, ed. D. L. Klass, Institute of Gas Technology, Chicago.

Casler, M. D. & Jung. H.-J. (1999). "Selection and Evaluation of Smooth Bromegrass Clones with Divergent Lignin or Etherified Ferulic Acid Concentration," *Crop Sci.* 39, 1866–73.

Chum, H. L., et al. (1994). "Status of the IEA Voluntary Standards Activity–Round Robins on Whole Wood and Lignins," pp. 1701–16 in *Advances in Thermochemical Biomass Conversion: Papers from the International Conference on Advances in Thermochemical Biomass Conversion Vol. 2*, ed. A. V. Bridgwater, Blackie Academic & Professional, Glasgow, U. K.

de la Rosa, L. B., et al. (1994). "Integrated Production of Ethanol Fuel and Protein from Coastal Bermudagrass," *Appl. Biochem. Biotechnol.* 45–46, 483–97.

Durot, N., F. Gaudard, and B. Kurek. 2003. "The Unmasking of Lignin Structures in Wheat Straw by Alkali," *Phytochem. 63*, 617–23.

Gollapalli, L. E., Dale, B. E. & D. M. Rivers. (2002). "Predicting Digestibility of Ammonia Fiber Explosion (AFEX)- Treated Rice Straw," *Appl. Biochem. Biotechnol. 98–100*, 23–35.

Greene, N., et al. (2004). Growing Energy: How Biofuels Can Help End America's Oil Dependence, Natural Resources Defense Council, New York (www.nrdc.org/air/energy/biofuels/biofuels.pdf).

Hames, B. R., et al. (2003). "Rapid Biomass Analysis: New Tools for Compositional Analysis of Corn Stover Feedstocks and Process Intermediates from Ethanol Production," *Proceedings of the Twenty-Fourth Symposium on Biotechnology for Fuels and Chemicals, 28 April–1 May, 2002, Gatlinburg, Tennessee*, ed. B. H. Davis et al., *Appl. Biochem. Biotechnol. 105-108*, 5–16.

Himmel, M. E., Ruth, M. F. & C. E. Wyman. (1999). "Cellulase for Commodity Products from Cellulosic Biomass," *Curr. Opin. Biotechnol. 10(4)*, 358–64.

Kelley, S. S., et al. (2004). "Rapid Analysis of the Chemical Composition of Agricultural Fibers Using Near Infrared Spectroscopy and Pyrolysis Molecular Beam Mass Spectrometry," *Biomass Bioenerg. 27(1)*, 77–88.

Kim, Y., et al. (2005). "Plug Flow Reactor for Continuous Hydrolysis of Glucans and Xylans from Pretreated Corn Fiber," *Energy Fuels 19(5)*, 2189–2200.

Lam, T.-B. T., et al. (1993). "The Relationship Between in Vitro Enzymatic Digestibility of Cell Walls of Wheat Internodes and Compositional Changes During Maturation," *Acta. Bot. Neerl. 42*, 175–85.

Lapierre, C., Jouin, D. & Monties. B. (1989). "On the Molecular Origins of the Alkali Solubility of Graminae Lignins," *Phytochem. 28*, 1401–3.

Laureano-Perez, L., et al. (2005). "Understanding Factors that Limit Enzymatic Hydrolysis of Biomass," *Appl. Biochem. Biotechnol. 121*, 1081–1100.

Lawoko, M., G. Henriksson, and G. Gellerstedt. 2005. "Structural Differences Between the Lignin-Carbohydrate Complexes Present in Wood and in Chemical Pulps," *Biomacromolecules 6*, 3467–73.

Lynd, L. R. (1996). "Overview and Evaluation of Fuel Ethanol from Cellulosic BIomass: Technology, Economics, the Environment, and Policy," *Annu. Rev. Energy Environ. 21*, 403–65.

Lynd, L. R., Elander, R. T. & Wyman, C. E. (1996). "Likely Features and Costs of Mature Biomass Ethanol Technology," *Appl. Biochem. Biotechnol. 57–58*, 741–61.

Moniruzzaman, M., et al. (1997). "Fermentation of Corn Fibre Sugars by an Engineered Xylose-Utilizing *Saccharomyces* Yeast Strain," *World J Microbiol. Biotechnol. 13*, 341–46.

Mosier, N., Ladisch, C. M. & Ladisch., M. R. (2002). "Characterization of Acid Catalytic Domains for Cellulose Hydrolysis and Glucose Degradation," *Biotechnol. Bioeng. 79(6)*, 610–18.

Mosier, N., et al. (2005). "Features of Promising Technologies for Pretreatment of Lignocellulosic Biomass," *Bioresour. Technol. 96(6)*, 673–86.

Sarikaya, A. & Ladisch, M. R. (1997). "An Unstructured Mathematical Model for Growth of *Pleurotus ostreatus* on Lignocellulosic Material in Solid-State Fermentation Systems," *Appl. Biochem. Biotechnol. 62*, 71–85.

Scobbie, L., et al. (1993). "The Newly Extended Maize Internode: A Model for the Study of Secondary Cell Wall Formation and Consequences for Digestibility," J Sci. Fd. *Agric. 61*, 217–25.

Sedlak, M. & Ho, N. W. Y. (2004). "Production of Ethanol from Cellulosic Biomass Hydrolysates Using Genetically Engineered *Saccharomyces* Yeast Capable of Cofermenting Glucose and Xylose," *Appl. Biochem. Biotechnol. 113–16*, 403–5.

Singh, R., et al. (2005). "Lignin-Carbohydrate Complexes from Sugarcane Bagasse: Preparation, Purification, and Characterization," *Carbohydr. Polym. 62*, 57–66.

"Standard Biomass Analytical Procedures: NREL Laboratory Analytical Procedures" (www.eere.energy.gov/biomass/analytical_procedures.html).

Wilson, J. R. & Mertens. D. R. (1995). "Cell Wall Accessibility and Cell Structure Limitations to Microbial Digestion of Forage," *Crop Sci. 35*, 251–59.

Wilson, J. R. & Hatfield, R. D. (1997). "Structural and Chemical Changes of Cell Wall Types During Stem Development: Consequences for Fibre Degradation by Rumen Microflora," *Aust. J. Agr. Res. 48(2)*, 165–80.

CELLULOSOME REFERENCES

Cellulosome Reviews

Bayer, E. A., et al. (2004). "The Cellulosomes: Multi-Enzyme Machines for Degradation of Plant Cell Wall Polysaccharides,"*Annu. Rev. Microbiol. 58*, 521–54.

Bayer, E. A., et al. (1998). "Cellulose, Cellulases, and Cellulosomes," *Curr. Opin. Struct. Biol. 8*, 548–57.

Demain, A. L., Newcomb, M. & J. H. Wu. (2005). "Cellulase, *Clostridia*, and Ethanol," *Microbiol. Mol. Biol. Rev. 69*, 124–54.

Doi, R. H. & Kosugi, A. (2004). "Cellulosomes: Plant-Cell-Wall-Degrading Enzyme Complexes." *Nat. Rev. Microbiol. 2*, 541–51.

Shoham, Y., Lamed, R. & Bayer, E. A. (1999. "The Cellulosome Concept as an Efficient Microbial Strategy for the Degradation of Insoluble Polysaccharides," *Trends Micro biol. 7*, 275–81.

Designer Cellulosomes

Bayer, E. A., Morag, E. & Lamed, R. (1994). "The Cellulosome—A Treasure Trove for Biotechnology," *Trends Biotechnol. 12*, 378–86.

Fierobe, H.-P., et al. (2005). "Action of Designer Cellulosomes on Homogeneous Versus Complex Substrates: Controlled Incorporation of Three Distinct Enzymes into a Defined Tri-Functional Scaffoldin," *J Biol. Chem. 280*, 16325–34.

Fierobe, H.-P., et al. (2002). "Degradation of Cellulose Substrates by Cellulosome Chimeras: Substrate Targeting Versus Proximity of Enzyme Components," *J Biol. Chem. 277*, 49621–30.

Fierobe, H.-P., et al. (2002). "Designer Nanosomes: Selective Engineering of Dockerin-Containing Enzymes into Chimeric Scaffoldins to Form Defined Nanoreactors," pp. 113–23 in *Carbohydrate Bioengineering: Interdisciplinary Approaches,* ed. T. T. Teeri et al., The Royal Society of Chemistry, Cambridge.

Fierobe, H.-P., et al. (2001). "Design and Production of Active Cellulosome Chimeras: Selective Incorporation of Dockerin-Containing Enzymes into Defined Functional Complexes," *J Biol. Chem. 276,* 21257–61.

In: Biological Barriers to Cellulosic Ethanol
Editor: Ernest V. Burkheisser

ISBN: 978-1-60692-203-3
© 2010 Nova Science Publishers, Inc.

Chapter 6

SUGAR FERMENTATION TO ETHANOL

United States Department of Energy

Fermentation of sugars by microbes is the most common method for converting sugars inherent within biomass feedstocks into liquid fuels such as ethanol. Bioconversion or biocatalysis is the use of microbes or enzymes to transform one material into another. The process is well established for some sugars, such as glucose from cornstarch, now a mature industry. Production of fuel ethanol from the mixture of sugars present in lignocellulosic biomass, however, remains challenging with many opportunities for improvement. More robust microorganisms are needed with higher rates of conversion and yield to allow process simplification through consolidating process steps. This development would reduce both capital and operating costs, which remain high by comparison with those of corn.

The growing U.S. industry that produces fuel ethanol from cornstarch has opportunities for incremental improvement and expansion. Processes for the bioconversion of lignocellulosic biomass must be developed to match the success in starch conversion (see sidebar, Starch: A Recent History of Bioconversion Success, this page). Technologies for converting cellulosic biomass into fuel ethanol already have been demonstrated at small scale and can be deployed immediately in pilot and demonstration plants. The challenge, with limiting factors of process complexity, nature of the feedstock, and limitations of current biocatalysts, remains the higher cost (see Figure. 1. The Goal of Biomass Conversion, p. 120). The discussion in this chapter will focus on process improvements that will reduce risk, capital investment, and operating costs. This emphasis is driven by the goal to integrate and mutually enhance the programs in DOE's Office of the Biomass Program (OBP) and Office of Biological and Environmental Research (OBER) related to achieving the president's goal of a viable cellulosic ethanol industry.

Bioconversion must build on its historic potential strengths of high yield and specificity while carrying out multistep reactions at scales comparable to those of chemical conversions. Biology can be manipulated to produce many possible stoichiometric and thermodynamically favorable products (see Figure. 2. Examples of Possible Pathways to Convert Biomass to Biofuels, p. 121), but bioconversion must overcome the limitations of dilute products, slow reactions, and often-limited reaction conditions. For commodity products such as fuels, biologically mediated conversion represents a large fraction of costs and selling prices (unlike the pharmaceutical industry, where bioconversion costs are small) (Lynd, Wyman, and Gerngross 1999). Ultimately, goals in this roadmap seek to define and overcome the biological limitations for key conversion parameters of metabolic flux and product, thermal, and pH tolerances to develop a robust bioconversion process. Several chapters articulate practical advantages and some challenges of biocatalysis and biomass conversion. While most biological research has focused on systems relevant to basic knowledge or medical applications, it has provided a wide base of tools and knowledge for application to the bioconversion of biobased feedstocks.

STARCH: A RECENT HISTORY OF BIOCONVERSION SUCCESS

Biotechnology has a track record of displacing thermochemical processing in the biomass starch industry. In the 1960s, virtually all starch (a sugar polymer in granules) was processed by acid and high temperatures. Inhibitory by-products and lower conversion rates resulted in a soluble starch solution that was lower in quality and yield when further fermented to ethanol. Development of specific thermostable high-productivity enzymes (e.g., alpha- amylase and glucoamylase) produced a higher-quality soluble starch, completely displacing the acid process by 1980. This new process has allowed technologies for producing ethanol from starch to continuously improve to the high yield and rate levels seen today in wet and dry corn mills. Other starch-conversion enzymes (e.g., glucose isomerase) have made possible another commodity product, high-fructose corn syrup, which is used in virtually all domestic sweetened beverages and many other products (www.genencor.com/wt/gcor/grain).

This discussion focuses on defining and prioritizing requirements for science and technology pathways that reach the maximal potential of biomass bioconversion. Results build on approaches developed in prior workshops (Scouten and Petersen 1999; Road- map for Biomass 2002). This chapter expands that focus in light of new biological research tools and understanding.The new biology will use such emerging technologies as proteomics, genomics, metabolomics, protein-complex characterization, imaging, modeling and simulation, and bioinformatics. This joint effort will further guide the development of new high-throughput (HTP) biological tools (e.g., screening, functional assays, and resequencing).

Some common themes arose during the workshop. (1) At present, we reaffirm recalcitrance of lignocellulosic biomass as a core issue, but portions of both the science and the conversion solution clearly are within the microbial world. (2) Understanding microorganisms will enable us to manipulate them so they can reach their maximal potential in human-designed processes. (3) The first thrust is to develop biocatalysts that will allow design and deployment of conversion processes that are less costly in operation and capital than current lignocellulose-to-ethanol conversion processes. (4) Another major thrust is to eliminate or combine separate processing steps by developing a "multitalented" robust microorganism. Research and development are addressing both strategies. (5) Even with molecular biology approaches, scientists create alterations (usually a single change) and observe the result. While experimental validation always is needed, new global genomics methods offer the potential for intelligently predicting the impact of multiple simultaneous changes.

Figure 1. The Goal of Biomass Conversion.
Securing cost- effective biofuels from biomass feedstocks requires moving biological technology from the laboratory (a. Microbial Cultures at Oak Ridge National Laboratory) through the pilot plant (b. National Renewable Energy Laboratory's Process Development Unit) to the full industrial biorefinery [c. Industrial Biorefinery in York County, Nebraska (Abengoa Bioenergy Corporation)].

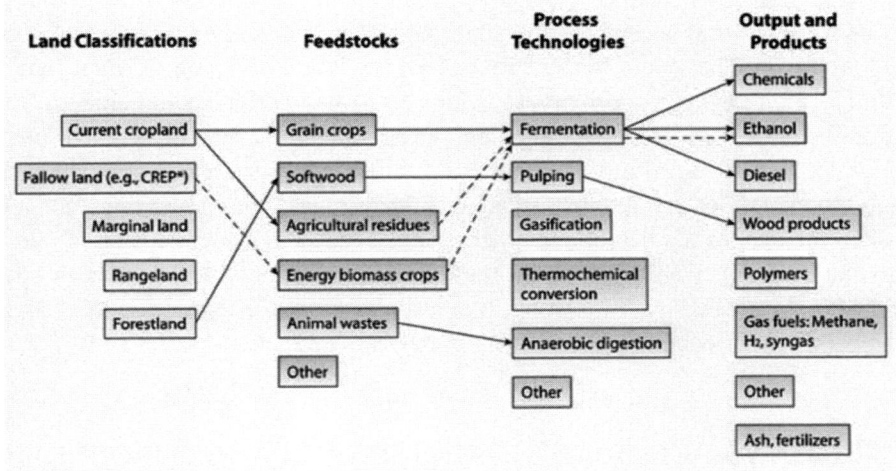

Source: B. Davison, Oak Ridge National Laboratory

Figure 2. Examples of Possible Pathways to Convert Biomass to Biofuels.
The dotted lines show examples of factors this roadmap can accelerate; solid lines indicate existing paths.

The new omic tools enable a deeper and more complete understanding of the microbial "state" and its physiology in its environment—enabling the probing of dynamics, regulation, flux, and function. Combining this understanding with the goal of improving microbial traits by manipulation will allow regulation of the microbe to achieve desired outcomes. Many traits or phenotypes, such as overall glycolysis rate or ethanol tolerance, will be multigenic. To identify further potential improvements, we especially need rapid methods to assess the state of microorganisms that have been engineered with new properties—either new process traits or industrial robustness. A first step is to analyze of how current industrial microbes have evolved through human selection from their progenitors to be better adapted to their process environments.

As stated, we need to achieve rapid analysis, modification, and understanding of the biocatalytic system to accelerate implementation of organisms for efficient bioconversion of sugars into ethanol. An array of basic microbial requirements includes full microbial system regulation and control, tools for rapid manipulation of novel microbes, and new microbial platforms. More practical requirements for biocatalysts include utilization of all sugars and a robust microorganism. The first may require deeper metabolic and regulatory understanding. The second requires an understanding of stress response and inhibition. It can be implemented by inserting all capabilities into one host or by using multiple microbial species with unique, complementary capabilities in a controlled, stable mixed culture. To enable this research and development, certain microbial-specific enabling tools are discussed. Through a deeper understanding of the microbial system, new biocatalysts can be developed to reduce process cost and risk in developing a truly sustainable industry.

Three core biological barriers have been identified as high-priority research areas for improving current bioconversion processes: Optimizing microbial strains for ethanol production, developing advanced microorganisms for process simplification, and creating tools and technologies to enhance the analysis, understanding, and use of microbial systems. We also consider several speculative, breakthrough opportunities offering novel approaches to biofuel production that could further reduce cost and risk in the more- distant future. These breakthrough, high-payoff opportunities include use of microbial communities rather than pure cultures for robust energy production, model-driven design of cellular biocatalytic systems, direct production of more energy rich fuels such as alkanes or long-chain alcohols, microbial production of up to 40% ethanol from biomass, and microbial conversion of biomass-derived syngas to ethanol and other products. Although such ideas as a pure in vitro multienzymatic system were considered, they seemed unlikely to compete with advantages microbes offer in producing, regulating, and using complex multistep carbon and energy metabolic pathways as commodities in the next 20 years.

OPTIMIZING MICROBIAL STRAINS FOR ETHANOL PRODUCTION: PUSHING THE LIMITS OF BIOLOGY

A major barrier in the efficient use of biomass-derived sugars is the lack of microbial biocatalysts that can grow and function optimally in challenging environments created by both biomass hydrolysis and cellular metabolism. The new tools of biology will facilitate the development of these advanced biocatalysts. Problems include inhibition by deleterious products formed during biomass hydrolysis, yields limited by accumulation of alternative products, unnecessary microbial growth, and suboptimal specific productivity resulting from various limitations in the ethanol biosynthetic pathway and a mismatch in conditions with the hydrolysis enzymes. Another challenge is that inhibition by the main fermentation product (ethanol) results in low alcohol concentration (titer). These problems contribute to the cost of lignocellulosic ethanol by increasing capital expenditure, reducing product yields, and increasing water volumes that must be handled as part of relatively dilute product streams. The research objective is to mitigate these limitations through concerted application of emerging tools for systems biology, working with principles from metabolic engineering and synthetic biology, and using evolutionary approaches combined with quantitative evaluation of candidate high-producing strains.

To foster an industry based on biomass sugars, process parameters must be comparable to those of the cornstarch ethanol industry. Ultimately, the overall cellulosic process can compete with petroleum, whereas cornstarch processes alone cannot achieve the needed quantities. Current technology is based on cornstarch conversion to ethanol utilizing yeast. This process uses glucose as the carbon source and converts it at high yields (90%), high titers (10 to 14 wt %), and reasonable rates (1.5 to 2.5 g/L/h). Recombinant ethanologenic organisms (i.e., yeast, *E. coli*, and *Z. mobilis*) have been created to ferment both glucose and xylose, but they currently produce lower ethanol titers (5 to 6 wt % ethanol). Improvements in ethanol yields and tolerance are needed to increase rates of production (>1.0 g/L/h) from all sugar constituents of lignocellulosic biomass. One successful strategy for utilization of both hexose and pentose sugars takes known ethanologens like yeast and adds abilities to utilize pentose sugars. Another strategy takes mixed-sugar consumers like E. coli and replaces native fermentation pathways with those for ethanol production. Figure 3. Changes in Metabolism Brought about by Genetic Engineering, this page, shows an example of how the output of a microbe can be changed. As titers are increased, rates slow down and eventually cease at ~6 wt % ethanol, the upper limit for wild-type *E. coli*. By comparison, wild-type yeast and *Z. mobilis* can reach titers of >15% ethanol from cornstarch glucose but have failed to achieve these levels on pentose sugars (see section, Optimal Strains: Fermentative Production of 40% Ethanol from Biomass Sugars, p. 149).

Most methods of biomass pretreatment to produce hydrolysates also produce side products (e.g., acetate, furfural, and lignin) that are inhibitory to microorganisms. These inhibitory side products often significantly reduce the growth of biocatalysts, rates of sugar metabolism, and final ethanol titers. In all cases, the impact of hydrolysates on xylose metabolism is much greater than that of glucose. Research described here offers the potential to increase the robustness of ethanologenic biocatalysts that utilize all sugars (hexoses and pentoses) produced from biomass saccharification at rates and titers that match or exceed current glucose fermentations with yeast.

From the above analysis of present and target states regarding use of biomass hydrolysates for biofuel production, critical parameters needed for a cost-competitive process are clearly evident:

High yield with complete sugar utilization, minimal by-product formation, and minimal loss of carbon into cell mass.
Higher final ethanol titer.
Higher overall volumetric productivity, especially under high-solid conditions.
Tolerance to inhibitors present in hydrolysates.

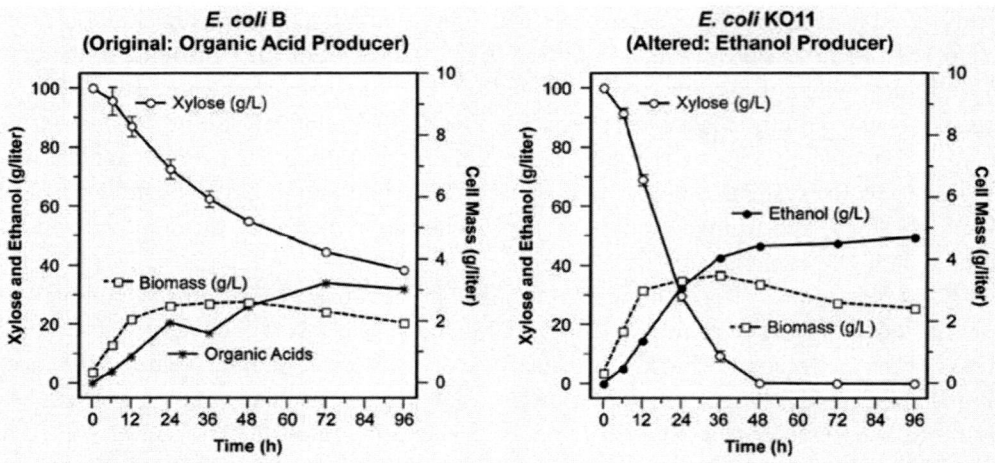

Source: L. Ingram, University of Florida. Based on data reported in H. Tao et al., "Engineering a Homo-Ethanol Pathway in Escherichia coli: Increased Glycolytic Flux and Levels of Expression of Glycolytic Genes During Xylose Fermentation," *J. Bacteriol.* 183, 2979–88 (2001).

Figure 3. Changes in Metabolism Brought About by Genetic Engineering.
The *E. coli* B strain, an organic acid producer, was altered to the *E. coli* strain KO11, an ethanol producer (ethanologen). The altered KO11 yielded 0.50 g ethanol per g xylose (10% xylose, pH 6.5, 35°C). In the graphs, biomass refers to the cell mass of *E. coli*.

Specifically, the following figures of merit are suggested for a biomass-to-ethanol process that will be cost-competitive relative to current cornstarch ethanol operations:

- Use of both hexoses and pentoses to produce ethanol at a yield greater than 95% of theoretical yield.
- Final ethanol titers in the range of 10 to 15 wt %.
- Overall volumetric productivity of 2 to 5 grams of ethanol per liter per hour.
- Ability to grow and metabolize effectively in minimal media or on actual hydrolysates (with only minerals as added nutrients).

To achieve the above targets, we must improve our ability to grow organisms in an inhibitory environment of high concentrations of sugars and other compounds, including ethanol. In addition, significant increases in flux through the sugar-to-ethanol metabolic pathway are needed. We present a roadmap below for meeting these objectives.

Science Challenges and Strategy

Key questions include:

- What are the implications of simultaneous vs sequential consumption of 5-carbon and 6-carbon sugars on cellular metabolism, flux, and regulation, especially when xylose metabolism has been engineered into ethanologens?

- What can allow more rapid and controlled alteration of microbes, especially regulatory controls and "adaptation" to novel inserted genes or deleted genes? This consideration applies also to known industrial microbes.
- What mechanisms control glycolytic flux, and what are their implications for cellular metabolism? For example, could the glycolytic pathway efficiently handle an excess of carbon flux in an organism engineered to rapidly consume a mixture of 5- and 6-C sugars (see Figure. 4. Recombinant Yeast, *S. cerevisiae*, with Xylose Metabolism Genes Added, p. 125)? A systems biology approach will allow insights into the molecular basis for these processes and development of predictive models to refine their design.
- What molecular mechanisms are used by cells to cope with such environmental challenges as high concentrations of sugars and ethanol and the presence of inhibitors from biomass hydrolysis?
- What genetic and physiological characteristics mediate evolution of wild-type organisms into robust laboratory or industrial strains, and which ones control their functional state in the process environment (see sidebar, Proteomic and Genomic Studies of Industrial Yeast Strains and Their Ethanol-Process Traits, p. 126)?

Utilizing a combination of metabolic engineering and systems biology techniques, two broad methods for developing more capable and more tolerant microbes and microbial communities are the recombinant industrial and native approaches.

- Recombinant industrial host approach: Insert key novel genes into known robust industrial hosts with established recombinant tools.
- Native host approach: Manipulate new microbes with some complex desirable capabilities to develop traits needed for a robust industrial organism and to eliminate unneeded pathways.

These methods require genetic understanding of the trait we wish to be added or preserved and robust tools for genetic manipulation. The subset of biochemical pathways potentially involved in glycolysis is complex (Figure. 5. Some Metabolic Pathways that Impact Glucose Fermentation to Ethanol, p. 128). Our goal is to pare this down to just what is essential for xylose and glucose use (Figure. 6. Desired Metabolic Pathways for a Glucose-Xylose Fermenting Ethanologen, p. 129). Both methods can have value; for example, either eliminate uneccessary pathways in *E. coli,* which has yielded strains that efficiently metabolize both xylose and glucose (and all other sugar constituents of biomass) to ethanol, or add xylose-fermenting pathways (and others) to ethanol-producing yeast. A number of methods and approaches support the two broad strategies. These and other goals will require certain enabling microbiological tools (see section, Enabling Microbiological Tools and Technologies That Must be Developed, p. 138).

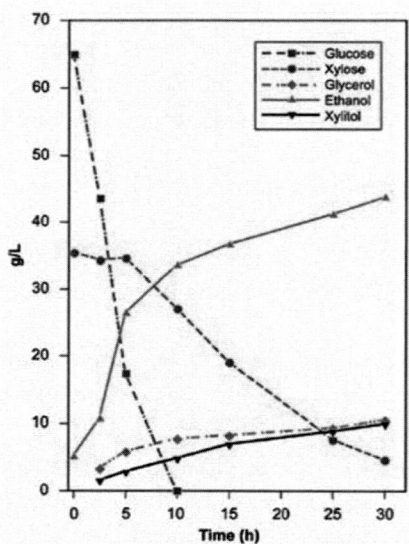

Source: M. Sedlak, H. J. Edenberg, and N. Ho, "DNA Microarray Analysis of the Expression of the Genes Encoding the Major Enzymes in Ethanol Production During Glucose and Xylose Cofermentation by Metabolically Engineered *Saccharomyces* Yeast," *Enzyme Microb. Technol.* 33, 19–28 (2003). Reprinted with permission from Elsevier.

Figure 4. Recombinant Yeast, *S. cerevisiae*, with Xylose Metabolism Genes Added.
Following rapid consumption of glucose (within 10 h), xylose is metabolized more slowly and less completely. Ideally, xylose should be used simultaneously with glucose and at the same rate, but the xylose is not totally consumed even after 30 h. Also note that yield is not optimal. Although ethanol is the most abundant product from glucose and xylose metabolism, small amounts of the metabolic by-products glycerol and xylitol also are produced.

Metabolic Engineering

Yield and productivity enhancement will be accomplished by applying *metabolic engineering* concepts and methods. *Yield maximization* is tantamount to by-product minimization, which is achieved by eliminating branches of competing pathways that lead to unwanted products. This usually is done by deleting genes encoding enzymes that catalyze competing reaction pathways. If such pathways are responsible for synthesis of metabolites essential for cell growth and function, downregulation of these genes may be preferable to complete gene knockout. In all cases, optimal balancing of enzymatic activities is critical for satisfactory function of the resulting engineered strain. Current molecular biological methods can be deployed successfully to this end, including specific gene knockout, gene amplification through promoter libraries or regulated (induced) promoters, and other methods combining gene knockout or downregulation and gene amplification. The ability to measure detailed cell behaviors and develop predictive models to refine their design will be critical to speed up and enhance these engineering efforts.

A related part of this work is analysis and regulation of cellular energetics. Careful alteration of growth, energy, and redox often is needed. Frequently, decoupling growth from production will increase yield.

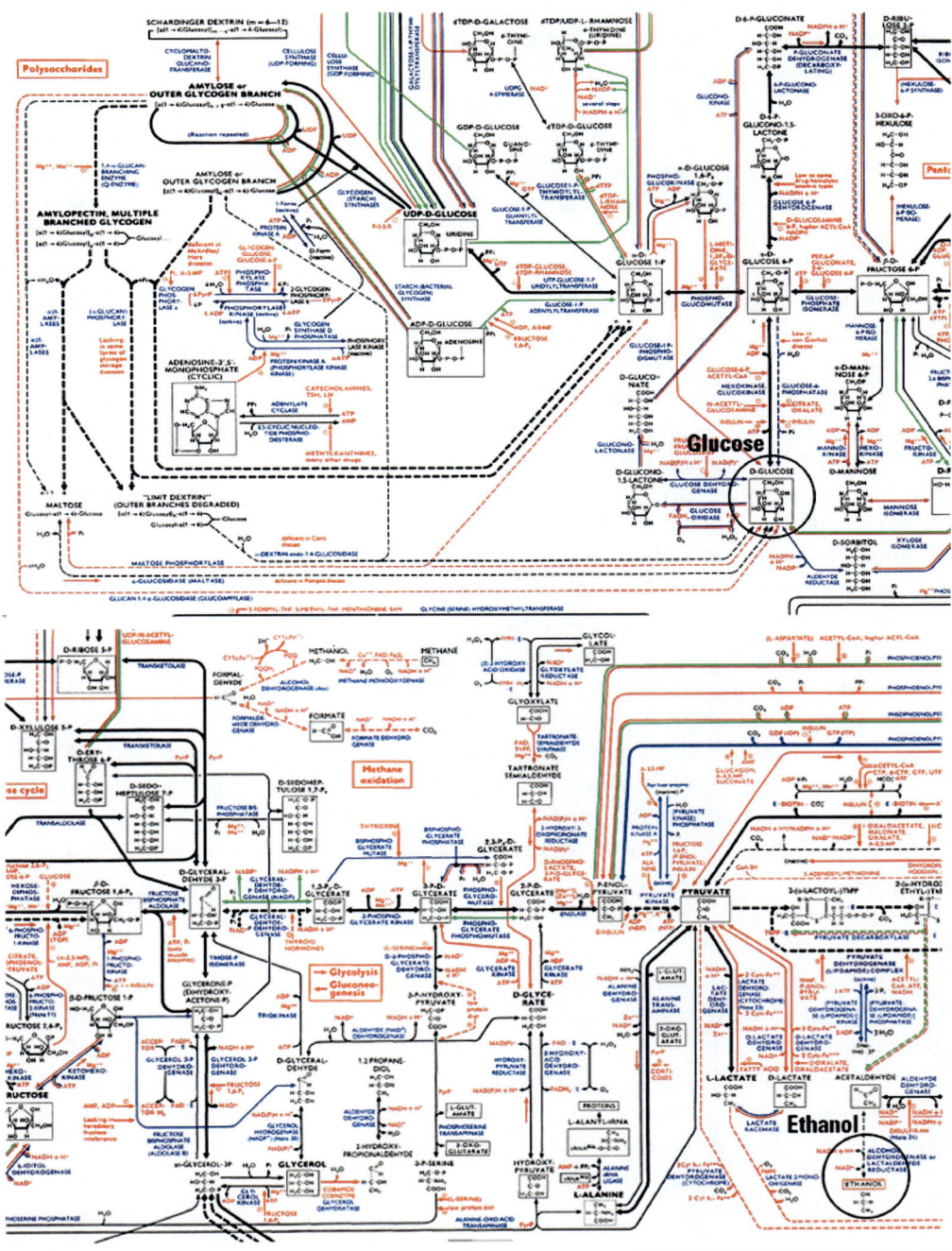

Figure 5. Some Metabolic Pathways that Impact Glucose Fermentation to Ethanol.
This pathway map demonstrates the complexity of even a simple, widely utilized, and relatively well-understood process such as glucose fermentation to ethanol. Glucose and ethanol are identified. [E. Gasteiger et al., "Expasy: The Proteomics Server for In-Depth Protein Knowledge and Analysis," *Nucleic Acids Res*. 31, 3784–88 (2003). Screenshot source: http://ca.expasy.org/cgi-bin/show_thumbnails.pl.]

Productivity maximization has been demonstrated in many applications of metabolic engineering with *E. coli* and yeast strains. Examples include 1,3 propanediol, amino acids such as lysine and threonine, biopolymer biosynthesis, precursors of pharmaceutical compounds, ethanol, and many others (see Figure. 7. A 3G Titer from Glucose, p. 130). These examples illustrate the feasibility of significant specific productivity enhancements by applying genetic controls, sometimes in combination with bioreactor controls. Improvements suggest that projected enhancements in specific cell productivity are entirely feasible and that the new technologies of systems biology can dramatically increase and accelerate results.

The first generation of specific productivity improvement will target enzymes important for the sugar-to-ethanol pathway. Stable isotopes will be used as tracers to map the metabolic fluxes of ethanol, including related pathways producing or consuming energy or redox metabolites (e.g., ATP or NADPH), and other key precursors for ethanol biosynthesis. Flux maps, together with transcriptional profiles, will be generated for control and mutant strains to identify enzymes controlling overall pathway flux. Gene amplification of rate-limiting steps will be used to overcome flux limitations. This is anticipated to be an iterative process, as new limitations are likely to arise as soon as one is removed by gene modulation. The goal will be to amplify flux of the *entire pathway* without adverse regulatory effects on the organism's growth or physiology. Again, balancing enzymatic activities, removing limiting steps, and pruning unwanted reactions—all supported by comprehensive analysis and modeling—will be deployed for this purpose.

In addition to specific pathway steps, remote genes with regulatory and other (often-unknown) functions impact pathway flux. Modulation of such genes has been found to influence significantly the biosynthesis rate of many products. Such genes will be found through inverse metabolic engineering, whereby libraries of endogenous and exogenous genes are expressed in the host strain and recombinants are selected on the basis of drastic improvements in the desirable phenotype (e.g., ethanol production and tolerance). Genes conferring these phenotypes can be sequenced and identified for expression in clean genetic backgrounds.

Recombinant Approach

Tolerance to inhibitors is a multigenic property. In the example systems given above, this trait is founded primarily on membrane fluidity and other membrane properties and functions. In general, efforts to improve microorganism tolerance by recombinant gene manipulation have been confounded by the limited ability to introduce *multiple* gene changes *simultaneously* in an organism. Development and use of a systems approach that allows multiple-gene or whole-pathway cell transformation are important milestones.

Evolutionary Engineering

A strategy for increasing ethanol tolerance or other traits could use *evolutionary engineering* concepts and methods. This strategy would allow the microbial process to evolve under the proper selective pressure (in this case, higher ethanol concentrations) to increasingly higher ethanol tolerances.

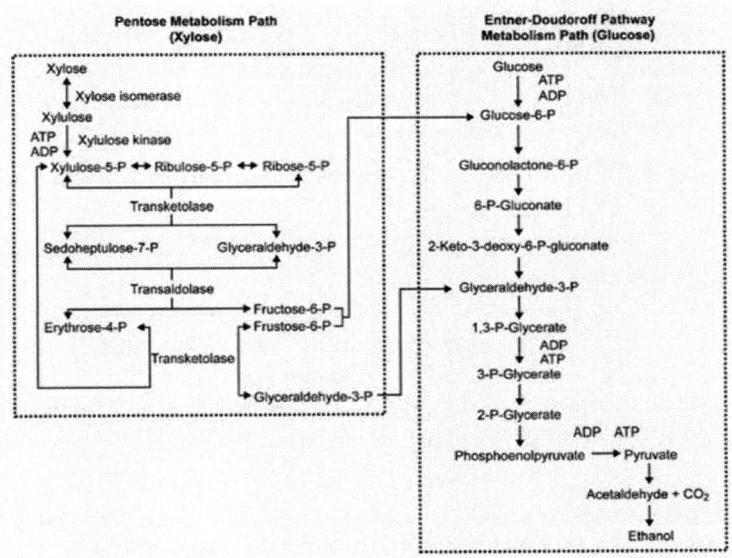

Figure 6. Desired Metabolic Pathways for a Glucose-Xylose Fermenting Ethanologen.
The goal is to genetically engineer an industrial organism that can metabolize both sugars. Pathways indicate the many involved genes that would have to be functional in such an organism.

Evolutionary engineering can be applied to the ethanol-producing organism as a whole or to specific proteins, in particular those with regulatory functions. In the latter case, evolutionary engineering emulates the methods of *directed evolution,* which has proven very successful in engineering protein mutants with specific desirable pharmaceutical, regulatory, or kinetic properties.

Accurate characterization of cell and protein mutants will be needed to allow an understand ing of principles for improving and rationally carrying out the designs. This task will require sequencing, large-scale binding experiments, transcriptional studies, proteomics, metabolomics, and physiological functional evaluation, all well suited for GTL capabilities described in more detail below.

Evidence is growing that methods of evolutionary engineering and directed evolution of regulatory proteins have the potential to achieve the targets of tolerance to ethanol and other inhibitory compounds. Recent studies with E. coli have increased that organism's ethanol tolerance by more than 50% while comparable increases also have been obtained for yeast at high ethanol (6%) and glucose (100g/L) concentrations. Applying systems biology methods and HTP technologies and computing will accelerate the process by revealing the genetic, molecular, and mechanistic impacts of evolutionary methods.

These methods also will be used to isolate fast-growing organisms. High growth rates and final biomass concentrations are imperative for achieving high-volumetric productivities, since the latter depend on fermentor biomass concentrations. More detailed investigation is needed on the effects of various biomass-hydrolysate compounds on cell *growth*, especially factors responsible for gradual reduction in specific ethanol productivity during fermentation as sugars are depleted and products, particularly ethanol and other inhibitors, accumulate.

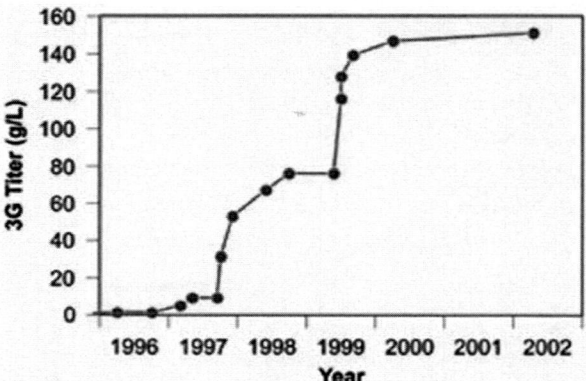

Source: Adapted from C. E. Nakamura and P. Soucaille, "Engineering *E. coli* for the Production of 1,3-Propanediol," presented at Metabolic Engineering IV, Tuscany, Italy, October 2002.

Figure 7. A 3G Titer from Glucose.
The graph shows DuPont-Genencor success in altering E. coli to maximize yield and titer of 3G (1,3, propanediol). Projects such as this could greatly benefit from the deeper systems biology understanding that GTL seeks.

PROTEOMIC AND GENOMIC STUDIES OF INDUSTRIAL YEAST STRAINS AND THEIR ETHANOL- PROCESS TRAITS: RAPIDLY FINDING THE GENETIC AND FUNCTIONAL BASES

Current industrial yeast strains have been isolated from wild yeast populations for many decades, selected for their capacity to produce ethanol under industrial settings. Understanding how these selected genotypes and phenotypes differ from undomesticated strains of yeast would help us to understand the type of changes needed to develop a robust ethanol producer. Understanding how cells cope with high-ethanol media concentrations is essential to improve fermentation yield and titer. Similar studies would be beneficial for other industrial organisms, such as Escherichia coli. Gaining insight about an organism's process of adaptation to an industrial biorefinery environment can help us intentionally replicate these changes (see Figure A. Importance of Adaptation for Robust Initial Strains, below). The strategy for studying industrial strains follows.

- Compare proteomic and genomic sequences of the most common yeast strains manufactured and sold for ethanol production with those of their ancestral parent strains.
- Compare proteomic and genomic sequences of evolved strains produced through metabolic engineering and metabolic evolution with those of their parental strains. Proteomic studies will be performed on samples taken from industrial fermentations. Genomic studies of strains from the same processes should reveal differences between industrial and laboratory strains that will provide fundamental information regarding multigenic traits essential for high metabolic activity, product tolerance, and adjustments to engineered changes in metabolism.

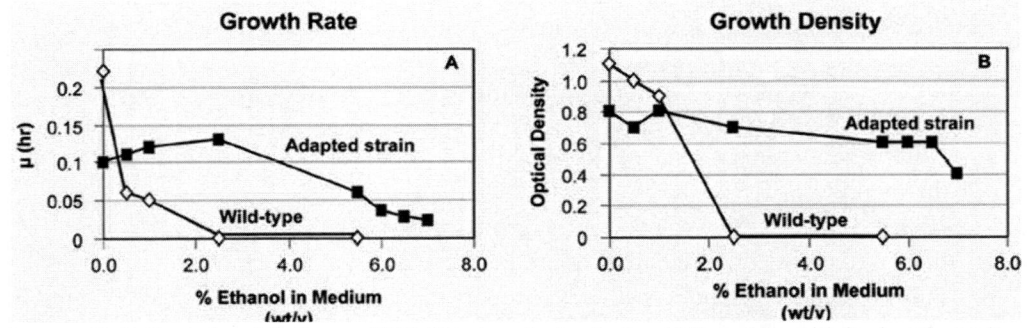

Source: H. J. Strobel and B. Lynn. 2004. "Proteomic Analysis of Ethanol Sensitivity in *Clostridium thermocellum*," presented in general meeting, American Society for Microbiology, New Orleans, La., May 23–27, 2004.

Figure A. Importance of Adaptation for Robust Industrial Strains.

Studying genomes of industrial yeast strains will help us understand common traits of effective ethanol-producing strains. Proteomic studies will reveal proteins generated under actual industrial production conditions. Complete mapping and reconstruction of the strain's networks will be needed for proper comparisons. Available modeling tools are being improved continuously. Proteomic analysis of membrane proteins is still a challenge and needs to be developed further to guarantee a more complete and meaningful analysis of samples.

Data generated through this effort will require full use of all tools available for systems biology and will stimulate hypothesis generation and testing by the academic community. The effect will be similar to those from metagenomic studies and community proteomics, in which huge amounts of data were made available and are being analyzed by many different groups around the world.

Technical Milestones

Within 5 years

- Mesophilic microbes demonstrated at scale that are capable of full utilization of all lignocellulosic sugars for reduced commercialization risk. This requires optimization of developed and partially developed strains.
- Increased strain tolerance to inhibitory hydrolysates and ethanol, with the ability to use all sugars, including mesophile and thermophile strains.
- Understanding of multigenic causes of industrial robustness.
- Candidate microbes such as thermophilic ethanologens compatible with desired cellulase enzyme optima. This allows process simplification to single-vessel fermentation with efficient use of all biomass-derived sugars (see section, Advanced Microorganisms for Process Simplification, p. 132).
- Development of coproducts.

Within 10 years

- Rapid tool adaptation and regulation of genetically engineered strains, including use of minimal media.
- Ability to engineer ethanol tolerance and robustness into new strains such as thermophiles.
- Higher-yield microbes via control of growth and energetics.
- Increased product titer to simplify product recovery and reduce water use.
- Full predictive metabolic pathway systems model for common industrial microbes, including regulation and identification of unknown genes (see section, Model-Driven Design of Cellular Biocatalytic Systems Using System Biology, p. 142).

Within 15 years

- Thermophillic microbes demonstrated at scale to enable simultaneous saccharification and fermentation.
- Further refinement of biofuel process and operation.

The Role of GTL Capabilities

As discussed, achieving these objectives will require the use of rational- combinatorial and evolutionary approaches to improve the properties of individual enzymes and organisms. To inform, enhance, and accelerate manipulation of new microbes, systems biology analyses (e.g., omic measurements, knockouts, tagging of proteins and complexes, visualization, and a bioinformatic core structure for data) will be applied. Once the novelties (e.g., pathways, proteins, products, traits, and complexes) are identified, additional genetic tools will move desired genes and traits into a known industrial host or further manipulate novel microbes into an industrial organism by adding gene traits. There are no consistent and rapid tools for these manipulations at present.

The capabilities listed below will play an important role in both cases.

Protein Production

A wide range of proteins (regulatory, catalytic, and structural) will be produced and characterized, and appropriate affinity reagents will be generated. Modified proteins also will be used to understand functional principles and for redesign. Examples include glycolytic proteins and alcohol dehydrogenases from other organisms or those evolved in the lab, structural proteins from high-tolerance organisms, or regulatory proteins with altered properties.

Molecular Machines

HTP methods to identify binding sites of global regulatory proteins and other aspects of membranes and membrane formation will be required. Specific protein complexes of interest are sugar transporters, solvent pumps, or other porins. These measurements will inform our understanding of, for example, the interaction or association of enzymes along the glycolytic pathway. The membrane could be studied as a machine to control inhibitory stress.

Proteomics

Although rational and evolutionary approaches are envisioned, a common component of both is the use of tools that allow quantitative cellular characterization at the systems level, including existing tools for global transcript, protein, and metabolite profiling. Additional HTP tools not currently available will be required to monitor key players that define the redox and cell energy state [e.g., ATP, GTP, NAD(P)H, NAD(P)]. Capabilities could include metabolic flux mapping, a major activity in understanding and manipulating cellular metabolism. The most efficient way to estimate in vivo metabolic fluxes is through labeling experiments. Specific needs include appropriate nuclear magnetic resonance (NMR) and mass spectroscopy (MS) instrumentation and stable isotopes for visualizing pentoses, hexoses, and cellulose. Intensive mathemathical and computational power is required to achieve the final goal of flux estimation. HTP technologies for global identification of genes that impact ethanol biosynthetic pathways are required to select cells capable of high ethanol production and other desired functions.

Cellular Systems

The ability to track key molecular species as they carry out their functions and create predictive models for systems processes will be critical for developing or enhancing cell properties.

DOE Joint Genome Institute

Sequencing and screening of metagenomic libraries for novel genes and processes and analyzing novel organisms will be carried out at DOE JGI. Exploiting microbial diversity by mining for novel pathways or organisms that make a step change in ethanol production could spur the production of other chemicals through fermentation.

ADVANCED MICROORGANISMS FOR PROCESS SIMPLIFICATION

Methods and technologies discussed above will be applied to consolidating process steps, which is widely recognized as a signature feature of mature technologies and has well-documented potential to provide leap-forward advances in low-cost processing technology. In light of the complexity of underlying cellular processes upon which such consolidation depends, fundamentally oriented work will be a highly valuable complement to mission-focused studies and can be expected to accelerate substantially the achievement of applied objectives.

Realizing the benefits of targeted consolidation opportunities requires understanding and manipulating many cellular traits, an approach much more fruitful at a systems level than at the individual gene level. As discussed previously, examples of such traits include transporters, control mechanisms, and pathways relevant to use of non-native substrates (e.g., 5- C sugars and cellulose), microbial inhibition (e.g., by pretreatment-generated inhibitors or ethanol), and the ability to function well in simple and inexpensive growth media.

Investigation of these traits provides an important way to apply and extend new systems biology tools to nonconventional host organisms such as thermophiles.

The current process has undergone many improvements in the last decade. In Figure. 4 of the Introduction, p. 14, the process cartoon illustrates pretreatment (probably dilute acid hydrolysis), followed by a detoxification and neutralization step, then separate fermentation of the soluble pentose sugars. Some biomass solids are used to make the cellulases, which then are added to the biomass solids to convert cellulose to glucose, followed by a separate glucose fermentation. This section discusses recent and ongoing developments to make a single microbe for cofermentation of hexose and pentose sugars (e.g., glucose and xylose).

Eliminating process steps may reduce capital and operating costs and allow other synergistic benefits. Some of these simplification steps are under limited active research. We focus here on three immediate consolidation opportunities:

- Elimination of a dedicated step to detoxify pretreatment hydrolysates before fermentation. These inhibitors can be by-products of the hydrolysis process and include acetate, furfurals, and other undetermined substances. Figure 8. Recombinant Yeast Cofermentation of Glucose and Xylose from Corn Stover Hydrolysate Without Detoxification (this page) shows the impact of these inhibitors. In process configurations under consideration (e.g., acid hydrolysis), such detoxification requires equipment (e.g., solid-liquid separation and tanks), added materials (e.g., base for overliming followed by acid for neutralization before fermentation), and added complexity. Obvious savings can be realized by developing improved biocatalysts not requiring the detoxification step. For detoxification elimination, research will support development of organisms having a high tolerance to pretreatment-generated inhibitors or those that detoxify these inhibitors (e.g., by consuming them) while preserving other desired fermentation properties. Some inhibitors have been identified, such as furfurals and acetate, but not all are known.
- Simultaneous saccharification and cofermentation (SSCF), in which hydrolysis is integrated with fermentation of both hexose and pentose sugars but with cellulase produced in a separate step. For example, development of thermophilic ethanol-producing organisms for use in SSCF could allow the consolidated process to run at higher temperatures, thus realizing significant savings by reducing cellulase requirements. Previous analyses (Svenson et al. 2001) have shown that a midterm strategy to produce ethanol from biomass would be to develop new strains capable of yielding ethanol at 50°C pH 6.0, the optimal conditions for saccharification enzymes generated today by the industry.
- Combining cellulase production, cellulose hydrolysis, and cofermentation of C-5 and C-6 sugars in a single step termed "consolidated bioprocessing" (CBP). Widely considered the ultimate low-cost configuration for cellulose hydrolysis and fermentation, CBP has been shown to offer large cost benefits relative to other process configurations in both near-term (Lynd, Elander, and Wyman 1996) and futuristic contexts (Lynd et al. 2005).

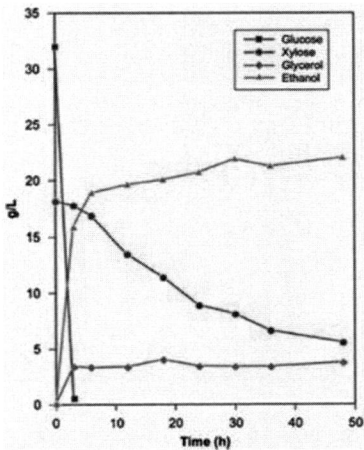

Source: M. Sedlak and N. Ho, "Production of Ethanol from Cellulosic Biomass Hydrolysates Using Genetically Engineered *Saccharomyces* Yeast Capable of Co-Fermenting Glucose and Xylose," *Appl. Biochem. and Biotechnol.* 114, 403–16 (2004) (also see Mosier et al. (2005)

Figure 8. Recombinant Yeast Cofermentation of Glucose and Xylose from Corn Stover Hydrolysate Without Detoxification.
Note slower xylose use and lower ethanol titer and yield than in Figure. 4. Corn stover hydrolysate was prepared by aqueous pretreatment followed by enzyme hydrolysis.

The goal is a process more like Figure. 5 of the Introduction, p. 15. Further simplifications can be envisioned beyond these examples. Unique challenges are the expression of multiple enzymes for cellulose and hemicellulose hydrolysis or the engineering of native cellulose-hydrolyzing organisms to produce ethanol. Selecting optimal enzyme targets for expression will require extensive screening and characterization of heterologous genes. Developing a unique enzyme suite capable of complete cellulose and hemicellulose hydrolysis will require insights into the plant cell-wall assembly and structure as well as new tools for cell-wall investigations.

Research must determine which aromatic hydrocarbon degradative pathways can solubilize lignin and how they can be integrated into a productive host for additional ethanol production. Fortunately, aromatic hydrocarbon-biodegradation pathways have been studied extensively over the past two decades, and many are known. Integrating necessary components into single hosts and channeling carbon to ethanol will be major challenges.

It will be a challenging goal to optimally achieve all the traits at one time. Expression, regulation, tolerance, growth, and metabolism must be designed and synchronized to function in the process. At present for this approach, we appear to be limited to anaerobic bacteria and not the aerobic fungi used to make current cellulases. We have limited knowledge and less ability to manipulate most of these bacteria. The cellulolytic bacteria also have some interesting differences, such as the cellulosome discussed in the biomass deconstruction chapter.

For all these consolidation opportunities, native or recombinant industrial strategies will be employed:

- The recombinant industrial strategy, engineering industrial organisms with high product yield and titer so cellulose or pentose sugars are used by virtue of heterologous enzyme expression.
- The native host strategy, engineering organisms with the native ability to use cellulose or pentose sugars to improve product-related properties (e.g., yield, titer) and process-related properties (e.g., resistance to toxic compounds).
- However, a third combined strategy is possible.
- Mixed culture conversion strategy to separately modify microbes to work on different parts of the substrates or pathways. This has been suggested but not well tested for cofermentation of pentoses and hexoses (see section, Microbial Communities for Robust Energy Production, p. 140).

As described before, the key difference is how challenging the complex trait is. Some cases like the elimination of lactic acid as a by-product might involve the deletion of a single gene; in others, the production of a complex extracellular cellulase or cellulosome may appear impossible at present. Delineating the genetic changes needed to confer resistance to toxic compounds generated in biomass processes is even more challenging. Researchers must balance the ease of manipulation against the trait's complexity in pursuing improvements of biocatalysts for industrial ethanol from lignocellulose.

For both SSCF and CBP, causes of cellulosic-biomass recalcitrance need to be understood not only with respect to enzymatic hydrolysis, in which enzymes act independently of cells (cellulose-enzyme complexes), but also to microbial hydrolysis, in which hydrolysis is mediated by cellulose enzyme-microbe complexes. Growing evidence shows that free-enzymatic and microbial hydrolyzes differ in substantial ways. Studies of recalcitrance in microbial cellulose hydrolysis will build on and complement, but not duplicate, investigation of enzymatic hydrolysis.

Further process simplifications can be considered. For example, development of robust, intrinsically stable pure or mixed microbial cultures could eliminate the need for costly sterilization. Alternative routes for process simplification also should be considered—such as gasifaction of the entire biomass followed by catalytic or biological conversion into fuels like ethanol (see section, An Alternative Route for Biomass to Ethanol, p. 152).

Science Challenges and Strategies for Process Simplification

The physiology or microbial state of modified organisms within the conversion process needs to be understood to help determine when simplification is helpful and what conditions must be achieved to make it effective. Part of this is regulation of native and modified pathways and traits, many of which appear to be multigenic, complex, poorly understood, and difficult to control. For all three consolidation opportunities, understanding the sensitivity of organism performance to growth-medium formulation would benefit from use of systems biology tools. Although separate processes are more cost-effective at times, simplification tends to win historically.

A. Elimination of Detoxification

- Fundamental mechanisms of toxicity and resistance.
- Evaluation of tolerance among a diversity of species and strains with and without opportunity for adaptation and evolution.
- Characterization and evaluation of detoxification mechanisms.

B. Simultaneous Saccharillcation and Cofermentation

- Fundamentals of fermentation in the presence of high solid concentration. The microbe-enzyme-solid interface should be analyzed.
- Understanding and reconciling factors responsible for differences in optimum conditions for cellulase function and fermentation of sugars to ethanol.

C. Consolidated Bioprocessing

- A key question is, How do microorganisms break down cellulose? How is breakdown in microbially attached cellulosome complexes different from enzymatic hydrolysis with added fungal enzymes? A significant number of fundamental issues needing to be addressed are over and above questions implicit in seeking to understand how enzymes hydrolyze cellulose. They include:

  Bioenergetics, substrate uptake, and metabolic control (including regulatory circuits) related to cellulose hydrolysis.

  Relative effectiveness of cellulose-enzyme-microbe complexes as compared to cellulose-enzyme complexes and the mechanistic basis for such differences and possible synergies.

  Extent to which products of microbial cellulose hydrolysis equilibrate or do not equilibrate with the bulk solution and the fraction of hydrolysis products that proceed from the cellulose surface directly to adherent cells.

  Features of cellulolytic microorganisms favored by natural selection and how selection can be harnessed for biotechnology (especially for the recombinant strategy).

  Documentation and understanding of the diversity of cellulose-utilizing organisms and strategies present in nature.

- How do microorganisms respond to cellular manipulations undertaken in the course of developing CBP-enabling microorganisms? Specific issues include:

  For the native strategy, how cells respond to changes in end-product profiles in terms of the cell's state (transcriptome, proteome, and metabolite profiles) as well as key properties of industrial interest (product tolerance and growth rate).

  For the recombinant strategy, understanding gained from recombinant cellulolytic microorganisms developed one feature at a time, including those in addition to hydrolytic enzymes (e.g., for substrate adhesion, substrate uptake, and metabolism). Such step-wise organism development provides an outstanding opportunity to advance applied goals and gain fundamental insights simultaneously.

D. Other Simplification Opportunities

- The development of intrinsically stable cultures could involve contamination-resistant thermophiles or acidophiles or techniques to control mixed microbial cultures (see section, Microbial Communities for Robust Energy Production, p. 140).
- The development of microbial growth-independent processes will reduce waste-treatment volumes of biosolids and allow better return of nutrients to the land as sustainable fertilizers.

Technical Milestones

Of the targeted consolidation opportunities, CBP is the most ambitious and probably will require the largest effort to achieve. Thus, we may well see substantial progress toward SSCF and detoxification elimination before CBP.

Key milestones associated with targeted consolidation opportunities can be pursued beneficially by complementary mission-oriented and fundamentals- focused research activities. These milestones include:

Within 5 years

- Improve hydrolysate-tolerant microbes.
- Achieve SSCF under desirable conditions (high rates, yield, and titer; solids concentration and industrial media).
- Functionally express heterologous cellulases in industrial hosts, including secretion at high levels and investigation of cell-surface expression.
- Conduct lab tests of modified initial CBP microbes.

Within 10 years

- Eliminate the detoxification step by developing organisms highly tolerant to inhibitors.
- Have the same response with undefined hydrolysates as with defined hydrolysates.
- Move to pilot demonstration of CBP.

Within 15 years

- Develop intrinsically stable cultures that do not require sterilization.
- Achieve CBP under desirable conditions (high rates, yield, and titer; solids concentration and industrial media), first on easily hydrolyzed model cellulosic substrates, then on pretreated cellulose.
- Develop methods to use or recycle all process streams such as inorganic nutrients, protein, biosolids, or coproduct carbon dioxide (see sidebar, Utilization of the Fermentation By-Product CO_2, p. 138).

The Role of GTL Capabilities

Protein Production
Protein production resources could be very useful in synthesizing enzymes and mixtures of enzymes as controls in experiments comparing enzymatic and microbial hydrolysis. These controls have the potential to be quite complex and thus demanding in terms of protein synthesis capability.

Molecular Machine Analysis
These resources can provide advanced analytical and computational science to study cell and cellulose interaction and particularly to gain insights into what is going on in the gap between an adhered cell and cellulose surface.

Proteomics
Proteomic capabilities can assist researchers seeking to understand system-level responses to metabolic manipulation in the course of developing microorganisms to achieve all three targeted consolidation opportunities as well as diagnosis and alleviation of metabolic bottlenecks and flux analysis.

Cellular Systems
These capabilities also can assist researchers in understanding system-level responses, removing bottlenecks, and conducting flux analysis (e.g., via metabolite analysis).

DOE Joint Genome Institute
DOE JGI can play a key role in sequencing genomes of new microorganisms with relevant features (e.g., ability to use C-5 sugars, resistance to pretreatment-generated inhibitors, and cellulose utilization), thus enabling virtually all lines of inquiry described in this chapter and in Crosscutting 21st Century Science, Technology and Infrastructure for a New Generation of Biofuel Research, p. 155.

UTILIZATION OF THE FERMENTATION BY-PRODUCT CO_2

Carbon dioxide is a major by-product of alcoholic fermentation by both yeast and bacteria. This relatively puregaseous stream requires no primary separation or enrichment step to concentrate the CO_2, which can be sequestered as part of the national climate-protection program or processed by biological or other means to useful coproducts. Technologies could be developed to produce value-added compounds that might provide income for financing ethanol biorefineries.

The fundamental challenge is how to supply chemical energy to use and reduce CO_2, produce useful compounds, and elucidate factors governing efficient use of CO_2.

ENABLING MICROBIOLOGICAL TOOLS AND TECHNOLOGIES THAT MUST BE DEVELOPED

Cellulosic biofuel research will use a broad range of powerful omic tools targeted for fuller development of the plant, enzyme, and microbial arena. However, some specific microbiological tools will need to be created to further understand and exploit microorganisms. These tools include analytical technologies and computational approaches and technologies for revealing the state of a microbial system, permitting assessment of perturbation effects on the system, and providing the information needed to construct useful models to guide engineering efforts.

The chapter, Crosscutting 21st Century Science, Technology and Infrastructure for a New Generation of Biofuel Research, p. 155, discusses in detail barriers in (1) gene-transfer methods and expression of genes in nonconventional host organisms; (2) tools for rapid analysis and modeling of cellular composition and physiological state; and (3) HTP screening methods for novel and evolved genes, enzymes, cells, and communities. Additional required tools include the following:

- Devices and requirements for preparing well-controlled microbial samples for omic analysis.

    Integrating biological studies into a whole-systems understanding is being made possible by new analytical techniques. Systems biology needs to be driven by an organism's biological context and its physiological state, which is linked tightly to its complete history, including that of culture. For experimentation in all aspects of work with omic tools, high-quality reproducible samples are paramount for subsequent analysis or purification.

    Controlled cultivation is the method to provide these samples. Cultivation also is that part of the experiment where knowledge of the biology is critical, and the quality of subsequent understanding is driven by the design of microorganism cultivation. The emphatic consensus of workshop attendees was that chemostat or continuous, stirred-tank reactors will provide the highest-quality biological samples for measuring multiple properties (omics) because they maintain environmental conditions at a steady state. For some omic techniques, batch operation will be chosen because of limitations in current cultivation technology and because of sample-number and amount requirements. Investigators must realize, however, that the increased amount or number comes at some cost to quality.

    Apparatus is needed to characterize mixed microbial populations. Cellulose is degraded in nature by mixed populations needing characterization beyond identification of its members. New tools and approaches are required to understand each population's contribution to cellulose degradation.

- Development of novel techniques and approaches is needed to carry out evolutionary biotechnology, especially for multigenic traits. New and more efficient methods to generate genetic and phenotypic variation in microbes are needed to increase capabilities for obtaining new phenotypes that require multiple simultaneous changes.

- Techniques and approaches are required for studying interactions between cellulolytic microbes and their substrates. A key step in cellulose degradation in nature is adhesion of microbes to the substrate. This dynamic process needs to be characterized with new and more quantitative and spatially, temporally, and chemically sensitive approaches and techniques.

Realization of targeted consolidation opportunities and advancement of relevant fundamentals will be served by a variety of crosscutting technologies and capabilities, including the following.

- Bioreactors—novel configurations, in some cases—to evaluate performance for consolidation opportunities and to test several key hypotheses.
- Evolutionary biotechnology to develop needed strains, including those with new capabilities. Application of these techniques will be advanced by miniature reactors and automated or controlled systems (continuous, semicontinuous, or serial culture) to maximize evolution rates. Some special consideration probably will be required to adapt evolutionary biotechnology to insoluble substrates.
- Improved gene-transfer and -expression technologies for unconventional host organisms and particularly for Gram-positive organisms, which have potential to be profoundly enabling with respect to all three consolidation targets.
- HTP screening for functional abilities and traits. This is needed for selection of the most improved strains, especially for nongrowth-associated functions. It also is needed to identify the function of *unknown* or *hypothetical* genes to allow better models and metabolic engineering.
- Tools to understand microbial mixed cultures in an industrial context.
- Scanning and other microscopic techniques, as well as experimental and computational approaches drawn from biofilm research, to characterize adhered cells.
- Systems biology tools (e.g., transcriptome, proteome, metabolome) to characterize intracellular events associated with targeted consolidation opportunities in both naturally occurring and engineered cells. This includes omic analysis for characterization of existing industrial microbes under production conditions to inform development of new biocatalysts.
- Quantitative modeling at the cellular level to test fundamental understanding and provide guidance for experimental work relevant to all three consolidation opportunities.
- Mesoscale molecular modeling to understand critical events occurring in the gap between cellulose and an adhered cell and its accompanying enzymes (see sidebar, The Cellulosome, p. 102).
- Models to confirm that consolidation and process simplification will be more cost-effective than separate optimized steps (see chapter, Bioprocess Systems Engineering and Economic Analysis, p. 181).

BREAKTHROUGH, HIGH-PAYOFF OPPORTUNITIES

Microbial Communities for Robust Energy Production

Most industrial bioconversions rely on pure cultures. All environmental bioconversions are based on mixed cultures or communities, with specialists "working" together in an apparently stable fashion. Examples of mixed communities capable of cellulolytic conversion are ruminant cultures and termite-gut cultures. Are there intrinsic biological reasons why communities could not be used for biofuel production? The fundamental question is, Are there stable self-regulating multiplex solutions for biofuels?

Microbial communities offer flexibility not present in monocultures, because the collective multiple metabolic pathways of microorganisms are activated as conditions demand. For example, microbial communities potentially could produce multiple forms of cellulase enzymes for use in industrial production of ethanol. In fact, multiple cellulases have been shown to be more effective than a single cellulase at processing complex and variable feedstocks. Mixed cultures tend also to be more robust, a characteristic needed for industrial-scale use.

Research Directions

This goal would require the ability to manipulate and use microbial communities to achieve industrial goals—not just the natural microbial goals of reproduction, survival, and net energy utilization. Current applications of mixed microbial cultures primarily are for waste treatment (i.e., anaerobic digestion or biofiltration). However, these technologies are poorly understood and exploit natural selection for survival. There are limited examples of products from mixed cultures in the food industry, but modern biotechnology has used only pure cultures for pharmaceuticals or for bioproducts such as ethanol. The first steps in applying microbial communities to biofuels are (1) characterize and understand existing cellulolytic microbial communities of microbes (e.g., ruminant, termite, and soil) and (2) develop techniques to understand and stabilize intentional mixed cultures. Research can elucidate detailed population interactions (e.g., both trophic and signaling) that stabilize the community. Support also is needed to evaluate robustness and population drift over time, since many mixed- culture operations will be continuous. Gaining a deeper understanding about community evolution will allow the use of selective pressure methods to evolve consortia with increased cellulose-processing efficiency. As an additional benefit, knowledge of mixed-culture dynamics may allow development of new methods to make pure cultures resistant to biological contamination.

Scientific Challenges and Opportunities

The most basic requirements are for mixed-culture identification and enumeration. Major science challenges are analysis and measurement of the mixed-culture "state." Most current omic analyses are predicated on knowledge of the gene sequence. Mixed cultures increase the challenge—for sequence, transcriptomics, and proteomics. The challenge increases geometrically for lower-number (<1%) representatives of the community. Single species existing as a population of clonal variants also adds complexity. Therefore, new techniques need to be developed and tested for sequence-based "metaomic" analysis. Also possible are nonsequencebased, metaomic techniques (e.g., transcript function–based microarrays) (Zhou et al. 2004). The metabolome is the only analysis that will not be significantly more difficult compared with that analysis in pure culture. One subset of mixed cultures is the many industrial strains that live as a group of clonal variants (a single-species population). Analysis of how these actual strains have adapted to their working environment could be helpful.

After basic analysis, enumeration, and quantitative meta-omics, the goal is to understand community structure. Signaling molecules are known to be important in many communities, so we need to identify and confirm these molecules and determine their importance. Then we can consider how to modify these signals to control the community. Regulatory and metabolic modeling of individual members, as well as the community, will be essential to deciphering the regulatory structure. Understanding the physical structure will require imaging technologies to acquire detailed visualization and specific labeling of individual species. This might require individual-species modification to express tagged marker proteins and then to reassemble the mixed culture. Computational models combining omics and biochemical and spatial variables will be critical to accomplishing this goal.

Reproducible samples and improved cultivation techniques in highly instrumented chemostats, for example, also will be required, especially when lignocellulosic solids are introduced. In this case, reproducible samples are especially needed. Mixed cultures may not be deterministic. Some evidence shows that the final state is highly variable. Previous work has shown that parallel enrichments from the same natural source each led to different populations after multiple serial transfers.

GIL Facilities and Capabilities

A major priority, as described in the GTL Roadmap, is to understand microbial communities. Capabilities being developed in the GTL program are ideally suited for developing industrial use of microbial communities. GTL will rely heavily on the DOE Joint Genome Institute (JGI) for sequencing and resulting annotation. Integrated proteomic capabilities will be useful for a wide range of omic analyses. Signaling molecules and tagged proteins and clones would be provided from protein production resources. For advanced community metabolic models, cellular systems capabilities are needed. When studying community interactions with a lignocellulosic medium, the National Renewable Energy Laboratory's Biomass Surface Characterization Laboratory will be valuable.

Model-Driven Design of Cellular Biocatalytic Systems Using Systems Biology

Systems-level modeling and simulation is the modern complement to the classical metabolic engineering approach utilizing all the GTL technologies and computing. Microbial organisms contain thousands of genes within their genomes. These genes code for all protein and enzyme components that operate and interact within the cell, but not all proteins are used under all conditions; rather, an estimated 25% are active under any given condition in a living cell. We currently understand very little about how all these expressed proteins interact and respond to each other to create cellular phenotypes that we can measure or try to establish. With improved understanding of such complex cellular systems, we should gain the ability to design them in an intelligent manner. Figure 9. Microbial Models for Providing New Insights, this page, indicates some of this power and complexity. Two general and complementary methods are taken into consideration: The *synthetic* route, which embraces de novo creation of genes, proteins, and pathways and the *nature-based* route, which uses existing suites of microorganisms and seeks to improve their properties via rational design. This is a qualitative step beyond recombinant and native strategies discussed in the Metabolic Engineering section, p. 127.

The challenge starts with the ability to characterize cellular networks and then moves toward establishing computer models that can be used to design them. These models would capture all aspects of a microbe's metabolic machinery—from primary pathways to their regulation and use— to achieve a cell's growth. Figure 5 illustrates part of this pathway complexity. Through these computational models, we could design and optimize existing organisms or, ultimately, create novel synthetic organisms. Specific to biomass-to-biofuels objectives, organisms engineered using these technologies potentially will consolidate the overall process and reduce unit operations (see Figure. 9. Microbial Models for Providing New Insights, this page).

Research Directions

This research would provide the ability to control and optimize a microbial transformation from carbohydrates to ethanol and related value-added coproducts in a selected microbe. Gaining complete control over cellular networks implies a capability for engineering and consistently performing transformation with the best available yields, rate, and titers. It also includes the ability to design the microbe intelligently by using computational technologies for determining consequences and optimal approaches to intervene and engineer within cellular networks. Furthermore, this model would enable us to assess the limits of a bioprocess's microbial biotransformation (e.g. the maximal productivity and rates achievable) and potentially to engineer entirely novel biotransformation pathways and systems.

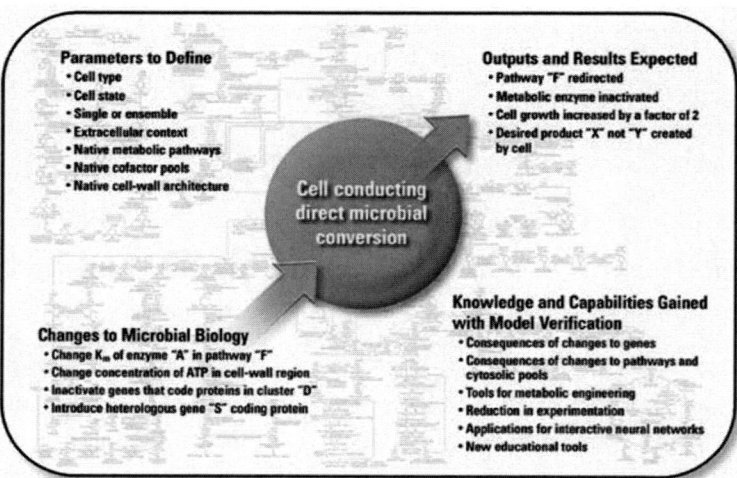

Figure 9. Microbial Models for Providing New Insights. Information gained in GTL systems biology research will enable metabolic engineering and modeling to enhance microbial characteristics. Using defined experimental parameters, the biology can be changed to perform desired new tasks. This will allow new biological system outputs, increase knowledge, and, ultimately, improve predictive models. [Source: M. Himmel, National Renewable Energy Laboratory]

Specific research directions would include the following.

Enumeration of Cellular Components, Interactions, and Related Phenotypes. Before any predictive computational models can be built, we need to generate the underlying data sets for relevant process conditions:

Identification of all proteins and enzymes participating in the metabolic pathways relevant to carbohydrate metabolism for cell growth, ethanol synthesis, and related by-products. This also would include characterization of key enzyme complexes relevant to cellulose degradation, carbohydrate transport, and respiratory mechanisms. Experimental technologies that may be useful include protein tagging, proteomics, and in vivo activity measurements.

Characterization of novel protein and gene function. About 30% of genes have no understood function, yet some of these unknown genes are thought to be involved in the microbial metabolic systems and stress responses under process conditions.

Identification and quantification of all metabolites present within the cell. Experimental approaches could involve NMR or MS.

Characterization of the cell's energetics under various relevant conditions. These would include measurements to characterize the stoichiometry of energy-transducing complexes, parameters such as the P/O and P/H+ ratios, and a cell's maintenance energy associated with cellular functions.

Characterization of transport mechanisms. In particular, this would focus on determining the components associated with transport of nutrients into and out of the cell as well as those of the mitochondria in eukaryotic organisms. Determining the stoichiometry of transporters and transport processes, as well as their kinetics and differential regulation, is envisioned.

Characterization of membrane composition for process tolerance (i.e., alcohol and toxin tolerance) and environmental and community interactions.

Elucidation of regulatory networks enabled by development of experimental approaches to identify protein-protein and protein-DNA interactions.

Localization data indicating where proteins are operating within the cellular and community space.

Knowledgebase to Develop Dynamics and Kinetics Modeling Techniques. One of the most attractive features of a model-driven, rational approach is its predictive capacity, requiring the inclusion of regulatory events at the genetic and metabolic levels. Creation of such models requires data sets, as well as development of HTP tools with capabilities beyond those currently available.

Acquisition of high-quality and dynamic omic data.

Development of HTP methods to identify binding sites of global regulatory proteins and other interactions.

Development of HTP tools to monitor key players that define the cell's redox and energy states [e.g., ATP, GTP, NAD(P)H, and NAD(P)].

HTP quantification of in vivo enzyme-activity metabolic fluxes.

Network Reconstruction. From data sets generated, we can develop the complete mapping and reconstruction of microbial networks and physiology related to the conversion of sugars to ethanol.

Automated techniques to integrate data sets and rapidly create reconstructed networks.

Integrated representation of metabolism, regulation, and energetics.

Approaches to account for the impact of spatial localization of proteins and enzyme complexes within integrated models.

Development of In Silico Analysis Tools. Methods are needed to interrogate and simulate the functioning of constructed networks to address key questions about microbial physiology. Any method should develop testable hypotheses that can be integrated with experimental studies. Methods should do the following:

Assist in network reconstruction, particularly in metabolic pathways and regulatory networks. These methods may involve new approaches that use artificial intelligence.

Interrogate mechanisms associated with toxic responses and tolerance to product and intermediate levels.

Assess physicochemical limitations of cellular systems and enzyme components to determine maximum achievable rates (e.g., identify rate-limiting steps, kinetic as well as diffusion limited).

Generate prospective designs of cellular networks by modifying and testing existing cellular systems.

Design systems de novo from cellular components.

5. **Design of Cellular Systems**. Designing engineered and synthetic organisms to convert carbohydrates to ethanol through the use of computational models and methods would include the following:

Dedicated transforming microbes with focused abilities to perform biotransformation and necessary supporting operations for the conversion of carbohydrates to ethanol.

Self-replicating synthetic microbes to support biofuel production under optimal conditions.

Novel pathways for producing biofuels and value-added coproducts from biomass could involve the generation of new enzymes and organisms through the use of evolutionary design concepts (e.g., directed evolution and adaptive evolution).

Scientific Challenges and Opportunities

To address this model-driven design goal, a number of broad scientific and conceptual challenges will need to be overcome, including the ability to make high-quality measurements of cellular components and states to simulate physiology and design networks with models generated from these data.

The Role of GTL Capabilities

Many GTL capabilities, either centralized or distributed, can be leveraged to aid in accomplishing these goals. Particular ones are noted below.

Protein Production

GTL capabilities will be used to characterize proteins by rapid isolation, production, and biochemical characterization in an HTP manner.

Molecular Machines

Molecular machine analysis will enable characterization of large complexes containing many active components of biotransformation networks.

Proteomics

HTP analysis of all proteins present in the cell, their relative abundance, spatial distribution, and interactions will be important to model development.

Cellular Systems

Ultimately, cellular systems analysis is about developing computational models of systems that can be used reliably to engineer microbes. Resources dedicated to the analysis and modeling of cellular systems can be used reliably by technologists to engineer microbes for biofuel production on an industrial scale.

Outcomes and Impacts

The GTL Roadmap describes scientific goals and milestones and the technology and computing needed to meet these research directions. These resources can be focused on the problem of engineering existing or synthetic organisms for biofuel production from biomass. This type of "rational design" and organism engineering has the potential to transform various stages of biomass conversion to biofuels consistent with goals of consolidating the overall process and reducing unit operations. The practical impact is in reducing the time required to modify a microorganism to perform as desired in an industrial setting.

Although the direct applied benefit will be in biomass-to-biofuel processes, technologies and methods derived from the ability to reliably engineer biological systems will have far-reaching impacts on basic and applied research across many sectors of biotechnology.

Direct Bioproduction of Energy-Rich Fuels

This breakthrough, high-payoff opportunity focuses on microbes for direct production of hydrophobic alternative fuels (i.e., alkanes, longer-chain alcohols, and fatty acids). This would overcome one limitation of nearly all bioconversions—they result in dilute aqueous mixtures. Typical industrial product concentrations are 100 to 150 g/L for ethanol and other such products as organic acids. This limitation imposes separation requirements that increase process and energy costs. New fermentation systems would be highly desirable to allow significant increases in product concentration, new types of products, and new processes for product recovery. Strong increases in efficiency also could be achieved by developing continuous processes.

Research Directions

Microorganisms produce a wide variety of potentially useful compounds but in relatively low amounts. Recently, because of expanded knowledge about the identity of genes for important pathways and mechanisms of pathway regulation, increasing the flux of microbially produced chemicals by up to six orders of magnitude (Martin et al. 2003) has been possible (from trace levels of primary products). A new opportunity is now offered to explore whether or not similar methods can be applied to developing modified microorganisms that secrete nontoxic molecules possibly useful for fuels. Examples may include alkanes, longer-chain alcohols, fatty acids (Voelker and Davies 1994), esters, and other types of molecules with low aqueous solubility that facilitate continuous product removal during fermentation. Advances in understanding how hydrophobic molecules are secreted by specialized cell types (Zaslavskaia et al. 2001) may facilitate the development of radically new production systems. The challenges described here for fermentation into hydrophobic fuels also would apply to potential photosynthetic systems.

Additionally, advances in systems biology and protein engineering may facilitate new approaches to the overall process of fermentation. For instance, developing chemical regulators of cellular processes such as cell division may be possible to allow cultures to be held in highly efficient steady states for prolonged periods. Such process controls may be synergistic with the development of novel product types not normally produced in high concentrations by microorganisms. For example, cocultures may possibly be used for directly combining alcohols and organics into ether or ester production. This would be an advantageous use of acetate released from biomass hydrolysis—taking it from a harmful by-product to a fuel cosubstrate.

Scientific Challenges and Opportunities

The explosion of sequence information resulting from GTL and other genome sequencing programs has greatly facilitated identification of genes for a wide variety of processes. This information expansion also has allowed the development of systems tools such as whole-genome DNA chips for measuring gene expression. A next-phase challenge is to bring that information and associated tools to bear on identifying entire pathways and cellular processes of relevance to biofuel production. Additionally, understanding how such pathways and processes are regulated is essential. New protein-production and proteomic tools envisioned for GTL will greatly facilitate the elucidation of pathways and their regulation. Important challenges are to understand how the permeability properties of membranes are controlled by composition and how the structure of membrane proteins such as transporters relates to function. Progress has been slow in elucidating membrane protein structure by conventional methods, requiring new approaches that may be addressed by GTL. Identification of microorganisms with high levels of resistance to biofuel compounds (but not necessarily to any production capabilities) could provide useful insights into strategies for improving fermentation efficiency.

The Role of GTL Capabilities

The full suite of GTL resources for genomics and systems tools will be essential in clarifying the underlying mechanisms associated with these and related problems. Examples of the types of contributions envisioned are listed below.

Protein Production

Protein production capabilities will enable elucidation of enzyme function in novel pathways for biofuel production; optimization of enzymes and transporters by protein engineering and evolution; and revelation of components for in vitro pathways. This could lead to development of novel chemical regulators of microbial cellular processes for use in industrial fermentation.

Molecular Machines

These resources will allow nanoscale interrogation of membrane interactions with biofuel compounds (e.g., using patterned membranes); identification of protein complexes; and mechanistic understanding of transporters involved in biofuel secretion. Development of nanoscale materials will facilitate product separations.

Proteomics

The proteomic approach involves biological-state omics for microbes under inhibitory stress; characterization of post-translational modifications of proteins that regulate enzymes or pathways for biofuel production; and analysis of biofuel exposure effects on microbial gene expression.

Cellular Systems

Cellular system capabilities include modeling of cellular carbon flux from uptake of biomass-derived sugars to secretion of finished biofuel compounds, systems engineering of batch and continuous fermentation for biofuel production, and modeling of protein structures in aqueous and nonaqueous environments.

DOE Joint Genome Institute

DOE JGI will characterize organisms with such useful properties as high productivity of or resistance to prospective biofuel compounds and will develop gene-expression interrogation systems.

Other Needs

Other needs (e.g., screening for new pathways and functions) include assessment of maximal redox balances (reduced fuel products yield more CO_2 in fermentation).

Outcomes and Impacts

If alternate fuels were made with higher fuel value (i.e., diesel, alkanes, lipids), both separations and life-cycle costs would be altered because these hydrophobic fuels would separate spontaneously from water. Fuel-density issues of ethanol also would be reduced. Additionally, transportation costs might be lowered because compounds such as alkanes would be significantly less corrosive than ethanol. These biofuels could be used more easily in the nation's current transportation infrastructure. If continuous fermentation with product removal were implemented, higher throughput would result in lower capital expenditures as well as costs associated with product dehydration, as in ethanol production.

Translation to Applications

DOE EERE would lead in pilot-scale tests of strains that produce novel biofuels and in developing fermentation processes based on new strains, products, and product-recovery processes. EERE would analyze the potential market and cost impacts for new and existing biofuels and then take the lead in separation technologies and in integrative separations. Additionally, EERE would carry out testing and possible engine-design modifications for new types of biofuels.

Optimal Strains: Fermentative Production of 40% Ethanol from Biomass Sugars

Current corn-to-ethanol processing plants typically produce titers of 10 to 14% w/w. Because of limits on biomass sugar concentrations, these ethanol levels are at least threefold higher than those produced from lignocellulose using current technology and biocatalysts. Tthe dilute product stream for lignocellulosic ethanol imposes a two- to threefold increase in fermentation volume per annual gallon of ethanol, with corresponding increases in pumps, nutrients, and management. Use of modern molecular tools to harvest the catalytic diversity of nature should facilitate construction of revolutionary biocatalysts that could increase ethanol titers up to 40% and lower needed investments in capital and operating costs for biomass-toethanol plants. This challenge is presented as an example of how systems biology may allow applications to exceed current biological limits. Strategies for obtaining organisms that produce and tolerate high-ethanol concentrations include:

- Engineering current ethanol producers to retain high metabolic activity.
- Engineering naturally ethanol-tolerant organisms to produce ethanol.
- Exploring native diversity of microorganisms to identify those that retain glycolytic and fermentative activity in the presence of high-ethanol titers.

Scientific Challenges and Opportunities

If achieved, production of fermentation broth containing 40% ethanol from biomass or starch sugars would revolutionize process designs. Incremental progress toward this goal would reduce dramatically the size of fermentation plants by decreasing fermentor capacity and associated pumps, nutrient cost, water usage, and waste-water treatment and recovery. For instance, doubling ethanol titers from 5 to 10% would reduce process water volume by 63% for equivalent ethanol production. Further doubling the titer from 10 to 20% would reduce the water needed by an additional 55%. For fermentation broth with a 40% titer, water usage would be only one-tenth the amount currently needed for the biomass-to-ethanol technology that produces 5% titer. At ethanol titers above 40%, viable alternatives to distillation could reduce energy costs associated with purification. However, distillation is a mature technology and its energy costs do not substantially limit the current process because waste heat is reused.

Improving the distillation process also would require developing better upstream processes or conversion technologies to provide highly effective sugar concentrations. A 30% biomass slurry can yield only a 15 to 20% sugar stream that results in just 7 to 10% ethanol. However, reflecting on this challenge illustrates how to further push biology beyond current limits.

Grain ethanol plants produce from 10 to 15% ethanol (Lynd, Wyman, and Gerngross 1999), reaching up to 20% ethanol when provided with very high substrate levels (Scouten and Petersen 1999). Even higher levels of ethanol (25%) are produced very slowly by sake yeasts at a low temperature. The most ethanol-tolerant microorganisms known, *Lactobacillus homohiochii* and *L. heterohiochii,* were isolated as spoilage bacteria in sake (Roadmap for Biomass 2002; Svenson et al. 2001; Lynd et al. 2005). The membrane lipids in these bacteria contain unusually long fatty chains that supposedly are an adaptation to growth in high ethanol. However, no systematic search for a class of extremophiles with resistance to ethanol has been reported.

Cells need not be viable to metabolize sugars to ethanol. At a biochemical level, ethanol production from sugars is a strongly exergonic reaction when coupled to ATP hydrolysis. Individual glycolytic enzymes in yeast and *Zymomonas mobilis* have been shown to function well in the presence of 20% ethanol (Zhou et al. 2004) and are progressively more inhibited at increasingly higher ethanol concentrations. In vitro disrupted cell preparations of both organisms have been shown to remain active and continue to produce ethanol even in the presence of 20% ethanol. Although organisms continuing to grow at ethanol concentrations above 30% are unlikely to be found, developing microorganisms that remain catalytically active and metabolize sugars to achieve very high levels of ethanol is quite plausible. For instance, over half the ethanol in commercial yeast fermentations is produced after growth has been inhibited by accumulated by-products. Formulating new biocatalysts for biomass presents a challenge and opportunity for engineering improvements to provide concentrated sugar feedstocks (Zhou et al. 2004).

Described below are several approaches for moving toward the goal of producing fermentation broth containing 40% ethanol, including the isolation of native novel microorganisms capable of growing or surviving in the presence of high levels of ethanol. Such organisms can be used as a platform from which to engineer ethanol production, if needed, and as a source of genes and enzymes to improve alcohol production in current ethanologenic biocatalysts. This work is expected to rely heavily on GTL resources for sequencing and transcriptome and proteome investigations, which will identify molecular requirements for ethanol tolerance during growth and for maintenance of active metabolism in the presence of high levels of ethanol.

Retaining the traits of current biocatalysts that do not disturb the metabolisms of all carbohydrate constituents of lignocellulose (hexoses, pentoses, and uronic acids) will be important. Additional genes for using various biomass carbohydrates and other components may be needed to provide high-substrate levels for high ethanol titers. These genes perhaps include hydrolases for cellulose and hemicellulose as well as uptake systems for solubilized products.

Other biomass components offer further opportunities to increase yield. Acetate levels equivalent to 10% of hemicellulose weight represent a potential source of oxidized substrate. Equivalent levels of solubilized lignins represent a source of reduced substrate, a potential electron donor to convert portions of both substrates into additional ethanol. Cometabolizing these substrates by engineering known genes and pathways from soil organisms could increase ethanol yield up to 5%. Acid-stable products represented by 4-O-methyl-glucuronoxylose and 4-O-methyl-glucuronoxylobiose in acid hydrolysates of hemicellulose currently are not metabolized by any ethanologenic biocatalysts. These recalcitrant products typically are not measured by high-performance liquid chromatography analysis and can represent up to 10% of total carbohydrate. Discovery of new genes and organisms to metabolize these saccharides and incorporate them into biocatalysts could provide a further incremental increase in ethanol yield with no increase in capital or operating costs. Together, the more complete use of all solubilized components from lignocellulose and the increase in ethanol titers would reduce dramatically the size of future ethanol plants and the amount of water use.

Research Directions

The burgeoning amount of sequence information and GTL's current ability to rapidly determine genome sequences for new alcohol-resistant organisms from nature provide an excellent opportunity. We can identify the genes and functions required for growth, survival, and continued metabolism in the presence of high levels of ethanol. Transcriptome and proteomic analyses resources will greatly assist these studies.

Characterizing microbial membrane and wall structures, including lipids, proteins, and carbohydrates, represents a difficult analytical problem and a necessary challenge for future GTL capabilities. Envelope structure is presumed to represent a major determinant for continued metabolic cell activity in high-ethanol and other extreme environments. New tools are needed to facilitate design and modification of biocatalysts for many future processes.

Other yet-to-be-discovered opportunities include potential metabolic pumps for solvents and products, possibly evolved by cells to maintain low intracellular product concentrations. New analytical tools will be needed to investigate properties and functions of these biomachines.

Process improvements also will be needed to make available the high amounts of sugars needed to achieve these ethanol titers, leading to solid- state fermentation in the future.

An Alternative Route for Biomass to Ethanol: Microbial Conversion of Syngas

Biomass can be gasified to produce syngas (mostly a mixture of CO and H_2). Perhaps surprisingly, syngas has been shown to be converted by certain microbes into products including ethanol (Klasson et al. 1990; Gaddy 2000). These microbes are not well understood, but the process has been taken to small pilot scale. The attraction of this alternative approach to bioethanol is that the theoretical yield is quite high since all the biomass potentially is available as syngas for anaerobic fermentation. This gives theoretical yields greater than 130 gal per dry ton of biomass.

Background

Gasification is a combination of pyrolysis and combustion reactions for converting a solid material, such as biomass, to a gasified product (syngas). Gasification is a robust and traditional technology, yet not extensively implemented.

Biopower can use this syngas as a fuel for power production. Once sulfur compounds have been removed, this gas can be converted to other products through catalytic Fisher-Tropsch reactions at high temperatures and pressures. However, these precious-metal catalysts for gas-to-liquid conversion have been explored for over 50 years with incremental improvements. Biocatalysts for some conversion methods are relatively unstudied, operating in aqueous media with the syngas bubbled past at ambient temperature and pressures and representing a strong alternative to traditional catalysis.

Challenges

How do these biocatalysts carry out transformations that otherwise work only with precious metals at high temperature and pressure? Which enzymes and molecular machines allow these transformations? Can increased understanding of these protein structure-function relationships aid development of either better biocatalysts or insights to improved inorganic metal catalysts?

Trial and error experimentation has shown that process conditions and reactor design will shift the microorganisms to higher product yields. This is the fundamental and unexplored biological question: How does the regulation of the fermentation pathway allow these environmental shifts (e.g., pH, and medium composition) to induce higher yields?

Syngas Status in Industry

Bioengineering Resources, Inc. is a small company developing and soon to be demonstrating its pilot syngas bioethanol process (EERE 2005; BRI Energy 2006). The University of Oklahoma has assembled an integrated gasification and biology program; however, rates remain slow and are limited by the fundamental biology and mass transfer (Klasson et al. 1990).

CITED REFERENCES

BRI Energy. (2006). "Technology Summary" (www.brienergy.com/pages/ process01.html).
EERE. 2005. "Synthesis Gas Fermentation" (www.eere.energy.gov/biomass/ synthesis_gas_fermentation.html).
Gaddy, J. L. (2000). "Biological Production of Ethanol from Waste Gases with *Clostridium ljungdahlii*," United States Patent 6,136,577.
Klasson, K. T., et al. (1990). "Biological Production of Liquid and Gaseous Fuels from Synthesis Gas," *Appl. Biotech. Biochem.* 24–25, 857–73.
Lynd, L. R., et al. (2005). "Consolidated Bioprocessing of Cellulosic Biomass: An Update," *Curr. Opin. Biotechnol.* 16, 577–83.
Lynd, L. R., Wyman, C. E. & Gerngross, T. U. (1999). "Biocommodity Engineering," *Biotechnol. Prog.* 15, 777–93.
Lynd, L. R., Elander, R. T. & Wyman, C. E. (1996). "Likely Features of Mature Biomass Ethanol Technology," *Appl. Biochem. Biotechnol.* 57–58, 741–61.
Martin, V. J. J., et al. (2003). "Engineering a Mevalonate Pathway in *Escherichia coli* for Production of Terpenoids," *Nat. Biotechnol.* 21, 796–802.
Mosier, N., et al. (2005). "Optimization of pH-Controlled Liquid Hot Water Pretreatment of Corn Stover," *Bioresour. Technol.* 96, 19 86–93.
Roadmap for Biomass Technologies in the United States. 2002. Biomass Research and Development Technical Advisory Committee (www.biomass. govtools.us/pdfs/FinalBiomassRoadmap.pdf).
Scouten, W. & Petersen, G. (1999). *New Biocatalysts: Essential Tools for a Sustainable 21st Century Chemical Industry*, Council for Chemical Research (www.eere.energy.gov/biomass/pdfs/biocatalysis_roadmap.pdf).
Svenson, C. J., et al. (2001). "Strategies for Development of Thermophiles for Ethanol Production from Lignocellulosics," presented at 23rd Sympo-sium on Biotechnology for Fuels and Chemicals, Breckenridge, Colo., May 2001.

Voelker, T. A. & Davies, H. M. (1994). "Alteration of the Specificity and Regulation of Fatty Acid Synthesis of *Escherichia coli* by Expression of a Plant Medium-Chain Acyl-Acyl Carrier Protein Thioesterase," *j Bacteriol. 176*, 7320–27.

Zaslavskaia, L. A., et al. (2001). "Trophic Obligate Conversion of a Photoautotrophic Organism Through Metabolic Engineering," *Science 292*, 2073–75.

Zhou, J., et al. (2004). *Microbial Functional Genomics*, New York, John Wiley & Sons, Inc.

In: Biological Barriers to Cellulosic Ethanol
Editor: Ernest V. Burkheisser

ISBN: 978-1-60692-203-3
© 2010 Nova Science Publishers, Inc.

Chapter 7

CROSSCUTTING 21ST CENTURY SCIENCE, TECHNOLOGY, AND INFRASTRUCTURE FOR A NEW GENERATION OF BIOFUEL RESEARCH

United States Department of Energy

OPPORTUNITIES AND CHALLENGES

Efficiently and inexpensively producing ethanol or alternative products such as alkanes, fatty acids, and longer-chain alcohols from biomass will require significant advances in our understanding and capabilities in three major areas explored at this workshop: Feedstocks for Biofuels, Deconstructing Feedstocks to Sugars, and Sugar Fermentation to Ethanol. A systems-level approach to understanding and manipulating plants and microorganisms central to processing biomass into liquid fuels depends on obtaining and using detailed chemical and biochemical information on organism states and structures to build functional models that guide rational design and engineering. A systems-level understanding of model plants will facilitate rational improvement of plant cell-wall composition in crops dedicated to conversion into biofuels. New approaches and tools will be necessary to characterize definitively the detailed organizational structures of principal types of plant cellulose and their relative energies and interrelationships with such other structural components as lignins and noncellulosic polysaccharides.

Similarly, systems-level explorations are needed to determine genetic makeup and functional capabilities of such microbial communities as those involved in biomass decomposition and sugar fermentation. Emerging tools of systems biology—together with principles and approaches from metabolic engineering, synthetic biology, directed evolution, and evolutionary engineering—will help to overcome current obstacles to bioprocessing cellulosic feedstocks to ethanol. Increased emphasis must be placed on achieving a predictive understanding of plant and microbial biology including dynamics, regulation, flux, and function, with the ultimate goal of rational design to improve traits of bioprocessing microorganisms and plant feedstocks. For example, a major barrier in efficient use of biomass-derived sugars is the lack of microorganisms that can grow and function optimally in the challenging environment created through biomass pretreatment, hydrolysis, and cellular metabolism. A milestone in surmounting this barrier is to identify and understand molecular mechanisms used by cells to cope with such environmental challenges as high sugar and ethanol concentrations and the presence of inhibitors from biomass pretreatment and hydrolysis. Because resistance mechanisms typically involve complex subsystems and multiple genes, systems biology is needed for rationally engineering microbes to overcome these limitations. Ultimately, the systems biology focus of the Genomics: GTL program (GTL) has the potential to enable consolidation of the overall bioconversion process and reduce the number of biorefinery operations.

Powerful new tools and methods, including high-throughput analytical and imaging technologies and data-handling infrastructure, also will be needed to obtain, manage, and integrate information into models. Related tasks involve sequencing model crop-plant genomes, understanding the structure and function of biomass deconstruction enzymes, and modeling metabolic and regulatory networks of microbial systems involved in bioconversion. Some of these technologies, which either exist now or are envisioned in the GTL Roadmap, are planned components of the proposed GTL capability investments; others will need to be enhanced or developed for specific research outlined in this report.

ANALYTICAL TOOLS TO MEET THE CHALLENGES OF BIOFUEL RESEARCH

Systems for effectively converting biomass to liquid fuels will require a wide array of innovative analytical capabilities to facilitate fundamental and applied science. These capabilities encompass new methods for rapid and sensitive analysis of biomass polymers and subunits and high-throughput characterization of plant cell walls and microbial populations catalyzing bioconversion reactions. Analytical methods should provide detailed information on chemical moieties, chemical bonds, and conformation of plant cell-wall polymers; they should rapidly assess the state of microbes having new properties or growing under various defined environmental conditions. As discussed in preceding sections, a capability requirement is analysis of membrane components including lipids, proteins, and carbohydrates. These structures possibly represent major determinants for continued microbial activity in high-ethanol and other extreme environments. Membrane analytics are a challenging problem but an essential capability. New tools are needed to facilitate biocatalyst design and modification for many future processes, and technologies, methods, and computational and informational tools will be established to support core systems biology. Such tools include genomics, transcriptomics, proteomics, glycomics and lignomics, and fluxomics as described below; imaging technologies at various lengths and time scales; structure-characterization techniques; biomass characterization; cultivation; and accompanying modeling and simulation capabilities.

Genomics

Capitalizing on potential biofuel production from cellulosic biomass requires the continued commitment of DOE's Joint Genome Institute to sequence various organisms that contribute to the process. These organisms include crop plants, industrial yeast strains, and microbial communities involved in biomass decomposition and soil productivity. Genetic blueprints provided by DNA sequences will allow the use of systems biology for rational design and process consolidation to increase production of biomass crops and optimize biofuel conversion processes. Support will be required for sequence assembly, optical mapping, and other techniques to place assembled contigs from a plant, microbe, or community on a physical map. Comparative genome analyses will infer similarities or differences in regard to other organisms, and expressed sequence tag and other libraries will facilitate genetic and molecular analysis of feedstock plants.

Genomic information will enable identification and molecular characterization of the diversity and activity of relevant biomass deconstruction enzymes from such new sources as fungi, bacteria, and uncultured microbes. Genomic analysis will open the door to use of genetic resources present in microbial communities specializing in lignocellulose degradation. Genomic projects are envisioned in the following general areas.

Feedstock Plants

This report outlines the importance of new and improved sources of feedstock biomass for conversion to liquid fuels. Considering project scale, soil and climate differences across the United States, and the desire to maintain biological diversity on agricultural lands, investigators recognize that a variety of biomass crops must be used. Grasses potentially are a prime source of biomass, so determining the DNA sequence of one or more grasses is a high priority. The genetic blueprint will assist in short-term development of grass-breeding strategies and the use of longer-term systems biology methods to realize the plant's potential as a source of biomass. Some plant systems to be studied include *Brachypodium* (slated for sequencing at DOE JGI) and *Populus* (already sequenced).

Soil Microbial Communities

Conversion into a high-value commodity is likely to alter the amount and composition of postharvest plant material and residues returned to the soil. Sustainability of biofuel technologies requires a scientific assessment of production practices and their influence on soil quality, including impacts on microbial communities and the processes they catalyze. Given that the technology involves crops tailored to thrive in individual regions, a number of sites representing various ecosystems must be chosen for soil metagenomic analysis. In the short term, such analyses will contribute to an understanding of effects on soil sustainability. The data also will be critical in preventing unwanted long-term effects on soil sustainability and for assessing more-direct effects of soil microbes on plant growth.

Fermentation and Biomass Decay Communities

Biomass conversion to sugars and biofuels requires optimizing microbial breakdown of structural sugars and fermentation of complex sugar mixtures. A number of microbial communities have evolved over millions of years to maximize and coordinate these capabilities, with several of the better-studied ones associated with ruminants and the hindgut of termites. Thus, a reasonable number of model fermentative communities that can degrade lignocellulose are critical targets for metagenomic analysis. In the short term, new insights into activities present or coordinated among members of these well-established communities will result. This information potentially can be used to design second-generation biofuel systems composed of microbial communities or consortia considered more robust and diverse in regard to environmental conditions they can withstand and types of substrates they can use.

Increased Production Systems

Adaptive evolution of microorganisms using selective pressure in fermentors or chemostats has enormous potential to select for traits that benefit biofuel production. To understand mutations that give rise to desired traits, determining the molecular basis of such changes is critical; we need to sequence industrialized strains and compare them to their progenitors. In the long term, this information is vital in designing new microbial systems that increase bioconversion efficiency and lower biofuel cost (see sidebar, Proteomic and Genomic Studies of Industrial Yeast Strains and Their Ethanol-Process Traits, p. 126).

Transcriptomics: High-Throughput Expression Analyses

Measuring RNA expression (transcriptomics) provides insight into genes expressed under specific conditions and helps to define the full set of cell processes initiated for coordinated molecular response. This information can be used to elucidate gene regulatory networks and evaluate models of cellular metabolism. Current technologies include microarray-based approaches applicable to the study of plants and homogeneous microbial populations in well-mixed systems. Custom microarrays targeted to new plant varieties or microorganisms are critically needed. In heterogeneous systems such as those in mixed microbial populations, current microarray-based approaches are less useful because they provide only an average gene-expression profile across an entire population. Therefore, single-cell gene-expression methods applicable to diverse cell types are required.

Proteomics

Although carbohydrates are biomass conversion's primary substrate, proteins are workhorse catalysts responsible for constructing biomass-forming polymers, depolymerizing biomass before fermentation, and converting sugars to desired end products such as ethanol. To interrogate these various processes, identifying and quantifying proteins and protein complexes for various plant and microbial systems and subsystems are vital. The genome's information content is relatively static, but proteins produced and molecular machines assembled for specific purposes are dynamic, intricate, and adaptive. All proteins encoded in the genome are collectively termed the "proteome." The cell, however, does not generate all these proteins at once; rather, the particular set produced in response to a specific condition is precisely regulated, both spatially and temporally, to carry out a process or phase of cellular development.

Proteomics can be used to explore a microbe's protein-expression profile under various environmental conditions as the basis for identifying protein function and understanding the complex network of processes facilitated by multiprotein molecular machines. Identifying the suite of proteins involved in construction and breakdown of cell-wall polymers is important in understanding biomass conversion to biofuels. In addition to analysis of intracellular proteins, all degradation enzymes and other proteins secreted by biomass-degrading microbes need to be characterized because they probably are key to the depolymerization process. Membrane protein systems are particularly important and need to be improved because they control cellulose production, cellulolytic enzyme excretion, and fermentable sugar transport.

To facilitate mass spectroscopy (MS) detection and quantitation of proteins in complex mixtures, improvements in instrumentation and chemistry are needed to enhance protein and peptide ionization, increase sensitivity and mass resolution, quantify protein levels in complex mixtures, and widen the dynamic range of their detection. New information and bioinformatic tools are required to analyze proteomic data.

Metabolomics

Analysis of the cell's metabolite content has lagged behind transcript and protein profiling but is an equally important indicator of cellular physiology. The metabolite profile may be a better indicator of cell physiology because metabolite concentrations (and fluxes) occur sooner in response to changes in the extracellular environment than do gene expression and protein production. Also, metabolites are precursors to the cell's transcripts and proteins and all other cell macromolecules, and regulation of cellular processes may not always be reflected in transcript or protein profiles.

The difficulty in profiling metabolites derives from their structural heterogeneity and short lifetimes inside the cell. Proteins and RNA are each composed of a constrained or limited set of precursors (amino acids in proteins, nucleic acids in RNA). Differences lie in amino acid or nucleic acid sequences, enabling a single separation method to analyze the entire transcript or protein profile at once. Unlike RNAs and proteins, metabolite heterogeneity makes analysis nearly impossible with a single separation technique. Although such techniques (e.g., high-performance liquid chromatography, thin layer chromatography, gas chromatography, and capillary electrophoresis) can separate metabolite groups having common structures, very few reports on separation and analysis of a comprehensive set of cellular metabolites have been issued. Furthermore, metabolites tend to have much shorter (seconds or less) half-lives than do proteins (hours) or RNAs (minutes). Shorter half-lives make rapid metabolism sampling and quenching even more important than analysis of RNAs and proteins. Methods such as nuclear magnetic resonance (NMR) that can effectively integrate and store information about flux through particular pathways and potentially can be applied to living systems. Localized metabolite concentrations such as those within an organelle also need to be measured.

Another complication is that a metabolite's intracellular concentration says little about its importance in cellular physiology. Indeed, some of the most potent cellular-signaling molecules found recently exist at relatively low concentrations inside the cell, and some metabolites produced and consumed at the highest rates have relatively low intracellular concentrations. In the former case, extremely sensitive methods are necessary to measure these metabolites. In the latter case, the metabolic flux profile is more important than absolute metabolite concentrations (see section, Fluxomics, p. 161).

Metabolomic research will address fuel production, including methods to isolate, extract, and analyze labile metabolites such as those involved in cellular energy metabolism (e.g., ATP, GTP, NADP, and NADPH). Multidimensional separation methods are vital to analyzing plant or microbial-cell metabolites. New analytical techniques are required for continuous metabolite measurements, monitoring processes, and in vivo metabolite analyses.

Glycomics and Lignomics

As parallel concepts to transcriptomics, metabolomics, and proteomics described in the GTL Roadmap, biomass-to-biofuels research will use similar high-throughput and high-content analysis with compounds having low molecular weight and involvement in synthesis and degradation of plant cell-wall polymers. Glycomics (profiling materials related to structural polysaccharides) and lignomics (profiling materials related to lignin) are essential capabilities. New powerful analytical tools will provide information for systems-level understanding of biomass structure and chemistry and its role in biomass conversion to fuels and valuable chemicals.

Information from these analyses will help us understand native and modified pathways for synthesis of cell-wall polymers by tracking precursor consumption, generating and utilizing intermediate structures, and exploring their connection to plant cell-wall chemical composition and physical-chemical structure. A "toolbox" of options for plant breeders and crop scientists eventually will be available to improve feedstock substrates specifically for fuel production. New glycomic and lignomic tools also will be applicable directly to other plant-development needs.

Information obtained from lignomics and glycomics will elucidate substrate modification by tracking structural changes and concentration fluxes in saccharification products that may be linked to harsh pretreatments. Saccharification-product profiling also will guide selection of appropriate enzymatic cocktails by identifying specific chemical bonds and functional groups in solids and larger oligomeric products. The addition of enzymes tailored to known recalcitrant structures could significantly enhance product yields. Information on the nature of recalcitrant structures also can be used to guide plant breeding and pretreatment conditions. These approaches will provide insight into rate-limiting steps that arise in substrate- and enzyme-limited systems.

Accurate and robust glycomic and lignomic analytical tools will play an important role in fermentation research. Many lignin- and carbohydrate- derived compounds are fermentation inhibitors, and other materials may be converted to undesirable side-products. Tracking these materials from feedstock through products will provide valuable insights into possible improvements in all stages of biomass-to-biofuels conversion.

Several specific challenges persist in the application of high-sensitivity, high-throughput tools for lignin and carbohydrate analysis. Many robotic and automated tools used in conventional metabolomics research cannot accommodate the larger sample sizes required for representative sampling of biomass substrates. Multivariate analysis tools will need to be developed and validated for accurate quantification of complex mixtures of substrates, pretreatment catalysts, conversion enzymes, and microbes. Many analytical methods for supporting a systems approach to biomass conversion will require biomass standards not readily available from commercial sources. Obtaining small-molecule standards will require prep-scale synthesis and isolation of molecules of interest and the ability to modify (e.g., by isotopically labeling) isolated materials using advanced techniques of carbohydrate and natural-product organic synthesis.

Fluxomics

Use of microorganisms for liquid-fuel production from lignocellulosic biomass depends on fluxes of diverse substrates through complex networks of metabolic pathways. Quantifying metabolic fluxes in microorganisms allows identification of rate-limiting steps in a biosynthetic pathway that could be improved by genetic manipulation or by alterations in cultivation conditions. Substrate turnover by an enzyme in a metabolic pathway can be determined by measuring changes in isotopically labeled substrate levels over a specific period of time. Using isotopomer analysis, quantities of labeled substrates and metabolic products derived from them can be measured and compared. Similar approaches can be used to map metabolic sugar fluxes to ethanol and related pathways.

Simultaneously measuring the turnover of all intracellular metabolites often is difficult, if not impossible. Once flux for certain enzymatic reactions is measured in the laboratory, flux through other cellular pathways can be calculated using mathematical models of the entire metabolic network. By modeling a given organism's metabolism, scientists can quantify the effects of genetic manipulations or changes in growth conditions on the cell's entire metabolic network. Inputs to flux-based models are the set of potentially active metabolic reactions and measurements of the steady- state production rates of such metabolites as DNA, RNA, carbohydrates, fatty acids, and proteins. Improved estimates of cell fluxes can be obtained by feeding a labeled carbon source and using measured transformed fluxes as input to the model.

A range of techniques and methods will be needed to determine fluxes in metabolic pathways of microbial cells employed for biomass fermentations. These techniques include stable and radioactive isotope labeling and associated methods for estimating unmeasured fluxes from isotope distribution in metabolites and macromolecules. MS- and NMR-based methods will determine isotope distributions in cellular metabolites and macromolecules.

THE SUPER IMAGER

The potential is to create compound, multifunctional instruments that individually include many of the following capabilities:

- Mapping of molecular species such as RNA, proteins, machines, and metabolites through the use of fluorescent tags of various kinds
- Multiple excitation and detection wavelengths including both fluorescent and infrared absorption methods
- High-speed 3D imaging

- Nonlinear contrast imaging including second- and third-harmonic generation and coherent Raman scattering
- Lifetime mapping as sensitive probes of local environments
- Rotational correlation mapping for in situ analysis of protein structure and function
- Magnetic resonance imaging with 10-micron–scale analyses of metabolite concentrations and providing data on diffusion properties and local temperatures
- Acoustical imaging at micron-scale resolution of the system's physical parameters
- Atomic force microscopy (AFM) mapping of structures with added information provided by the controlled interaction of light and sharp metallic AFM tips to obtain optical resolutions of ~20 nm, one-tenth the diffraction limit
- High spatial resolution (nanometer scale) using X-ray and electron microscopies, including the use of special DOE facilities or perhaps the development of laboratory-based X-ray sources for imaging

Sources: *Report on the Imaging Workshop for the Genomes to Life Program,* Office of Science, U.S. Department of Energy, 2002 (www.doegenomestolife.org/technology/imaging/workshop2002/); GTL Roadmap, pp. 182–87

New analytical techniques are needed to quantify extracellular metabolites and quickly and easily determine biomass composition.

Enzyme Structure and Function

Developing and deploying improved enzymes and multienzyme complexes for biomass deconstruction and conversion will require more understanding and the production of suitable substrates as well as enzymes and their appropriate complexes. They will be used to achieve a mechanistic understanding of cellulose and cell- wall interactions with degrading enzymes.

Defining and Producing Substrates

Despite nearly 100 years of pretreatment research, detailed understanding of cellulosic biomass's fundamental physical and chemical features is still lacking. The plant cell wall—the substrate for degradative enzyme systems—is complex, containing various forms and quantities of cellulose, hemicellulose, and lignin. Chemical and physical pretreatment may alter biomass to make enzymatic feedstock digestion more difficult, so merging pretreatment and deconstruction steps would simplify processing. Critical to understanding any enzymatic reaction is having both a reliable assay and a defined substrate, thus allowing activities to be measured accurately. Constructing standardized substrates suitable for high-throughput assays is required to optimize these enzymes. Synthesis and characterization of model substrates require multiple technologies, first to define the substrates by employing multiple analytical and computational methods and then to synthesize the substrates. High-throughput methods are needed for compositional analysis and characterization of biomass substrates, once a family of basic reference structures has been identified.

Identifying Enzymes and Degradative Systems

Multiple classes of enzymes are required for biomass conversion to achieve maximum sugar yields including hemicellulases and ligninases. The rate-limiting step is making cellulose accessible for subsequent saccharification steps. Surveying and identifying suitable enzymes require many technologies listed in the GTL Road-map, including metagenomic analyses of model biomass-degrading communities, high-throughput protein expression, generation of affinity tags, reassembly of complexes, activity measurements, and biochemical and biophysical characterizations of enzymes and complexes. Degradative systems fall into two general classes: Individual enzymes working synergistically; and the cellulosome, a large, mega-Da complex normally attached directly to cells and found to date only in anaerobic bacteria. The cellulosome is a LEG O-like system with many interchangeable structural and degradative protein components deployed when specified for particular substrates. This system provides the basis for engineering to create cellulosomes optimized for particular substrates. The cellulosome can be detached from cells and functions, even under aerobic conditions. With appropriate modifications, the cellulosome has potential for enhanced stability and activity. Individual enzymes and those in cellulosomes might benefit from immobilization technologies designed to increase activity and stability. How membrane proteins such as sugar transporters interface with cellulosomes attached to cell walls is unknown (see sidebar, The Cellulosome, p. 102).

IMAGING TECHNOLOGIES

"Thought is impossible without an image." —Aristotle, 325 B.C.

Imaging technologies with wide applications will be critical to many of the research challenges identified in this report. Biomass deconstruction is a crucial step in conversion, yet relatively little is known about the detailed molecular structure of plant cell walls and how they are constructed from various components. To address this limitation, new and improved methods are needed to analyze plant cell-wall composition and structure at the nanometer scale. Organization of polymer components in biomass structures should be analyzed in three dimensions using noninvasive tools. Such new capabilities are anticipated to reveal key molecular processes occurring in real time during the full life cycle of cell-wall formation and maturation and during experiments to transform crop species into feedstocks suitable for bioconversion by optimizing cell- wall makeup. Many requirements generally have been anticipated by GTL and by capability development and planning as documented on pp. 182–87 in the GTL Roadmap (U.S. DOE 2005) and Imaging Report (U.S. DOE 2002).

Imaging needs for each specific research area are summarized below. They include advances in a wide range of imaging technologies using NMR, optical, X-ray, and electron-based methods as well as atomic force and scanning tunneling microscopies. Given differences in resolution, sensitivity, and selectivity, the full impact of these methods will be realized best when used in combinations, either within a single instrument (see sidebar, The Super Imager, p. 162) or parallel applications assembled and correlated through advanced image-management software.

SOME IMAGING TECHNOLOGIES RELEVANT TO FEEDSTOCK CHARACTERIZATION

Atomic Force Microscopy (AFM)

In AFM, a scanning-probe technique allows the direct study of surface using tapping probes, the tips of which project less than a micron. The dynamic behavior of surfaces and molecules often can be observed. A great advantage of scanning force microscopy over most other high- resolution techniques is its ability to operate in a liquid environment. High-resolution AFM images of a cellulose surface recently have been reported. Using AFM to support interpretation of pretreatment and enzyme action on biomass surfaces represents a tremendous opportunity.

Scanning Electron Microscopy (SEM)

SEM really is the backbone of traditional biological surface analysis. New developments in the design of SEM sample chambers and "optics" permit the analysis of samples containing some natural moisture, which is critical for biomass fractions. Also, new strategies for creating replicas of biological samples provide more versatility in analyzing proteins and microbial- cell surfaces.

Transmission Electron Microscopy (TEM)

Cryoelectron microscopy is the leading technique for high-resolution molecular machine structural-biology research, and TEM is the most common platform for this method. The technique makes the following special demands on electron microscopy: Ultrahigh and clean vacuum for contamination-free observation, stable low-drift cryotemperature holders, special functions to provide low-dose imaging conditions, and cooled slow- scan charge coupled display cameras for low-dose digital image recording.

Imaging Needs for Feedstock Research

Within 5 years, methods will be developed and deployed for chemical-specific imaging over a wide range of spatial scales (0.5 nm to 50 µm), with contrast methods and tags enabling many molecular components to be distinguished easily. Imaging of living (or never-dried) materials is perhaps just as critical as imaging harvested materials that have been dried, either for storage in a processing facility or for structure examination. These imaging capabilities should enable observation of cell-wall deconstruction and construction. New optical technologies and advances at synchrotron light sources will allow some imaging methods to function as high-throughput devices, producing hundreds to thousands of images per minute with a single instrument. Important possible consequences are dramatic reduction of cost and processing time per sample and greater community access to these technologies as part of a user facility. Second, enhanced imaging speed enables real-time observation of biological processes. Availability of real-time data is vital to building and validating models that allow a systems biology approach.

Temporal imaging of dynamic subcellular small molecules, including metabolites, is among advances expected to take longer development times. Improved molecule-specific imaging tags and a means for introducing them into cells without disrupting function are critical needs. The scale of various biomass conversions will require enhanced imaging for obtaining high-resolution, large-volume tomographic images as well as real-time imaging of living systems at high spatial resolution and for use in the field at lower spatial resolution. The critical aspect of enzyme engagement with biomass components, such as cellulase interfacing with cellulose, requires new capabilities for characterizing enzyme binding sites by atomic force microscopy, scanning electron microscopy, transmission electron microscopy, and electron spectroscopy for chemical analysis [see sidebar, Some Imaging Technologies Relevant to Feedstock Characterization, p. 163, and Image Analysis of Bioenergy Plant Cell Surfaces at the OBP Biomass Surface Characterization Lab (B SCL), p. 40]. Similarly, capabilities are needed to characterize interactions between microbial cells and their solid-phase substrates.

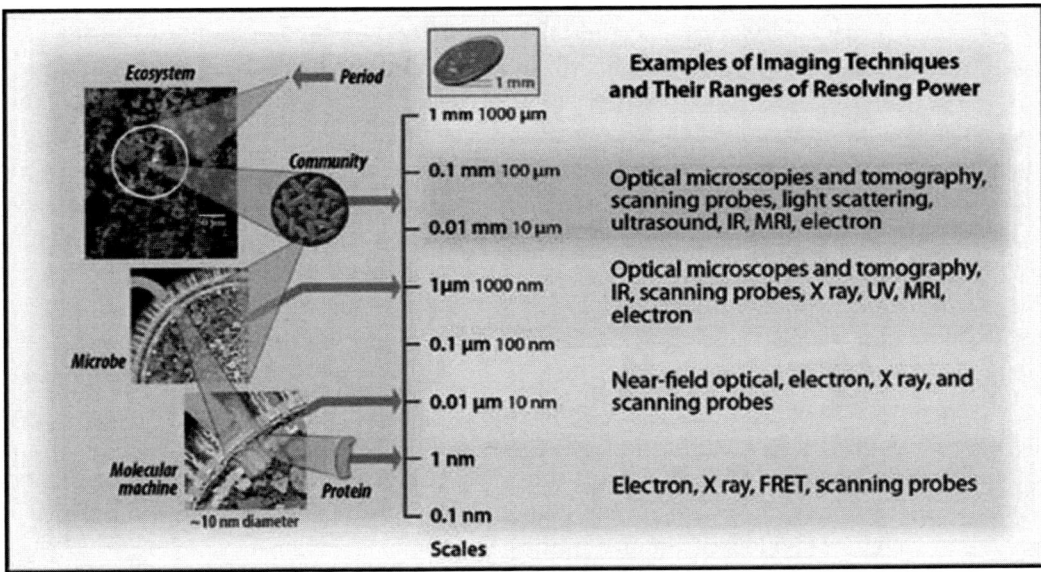

Source: GTL Roadmap, p. 176 (http://doegenomestolife.org/roadmap/).

Figure 1. Probing Microbial Communities.
Microbial communities and ecosystems must be probed at the environmental, community, cellular, subcellular, and molecular levels. The environmental structure of a community will be examined to define members and their locations, community dynamics, and structure- function links. Cells will be explored to detect and track both extra- and intercellular states and to determine the dynamics of molecules involved in intercellular communications. Probing must be done at the subcellular level to detect, localize, and track individual molecules. Preferably, measurements will be made in living systems over extended time scales and at the highest resolution. A number of techniques are emerging to address these demanding requirements; a brief listing is on the right side of the figure.

Imaging Needs for Microbial Communities in Deconstruction and Conversion of Biomass to Ethanol

Meeting imaging requirements for processing biomass to ethanol will draw on the above capabilities but will focus more on microbial-cell and -community imaging to aid in delineating sub- and extracellular organization of proteins, protein complexes, transporters, and metabolites. Appropriate tags and methodologies will be required for investigating multicellular interactions in mixed microbial populations (e.g., those observed in fermentation) and communities (e.g., those observed in colonization of decaying biomass) with solid substrates such as plant cell walls. Understanding these interactions will require information on amounts, types, and locations of secretion products, reactions they catalyze, and how these reactions and products impact cell viability. Imaging-based approaches are essential to achieve temporal and spatial resolution sufficient for understanding these processes (see Figure 1. Probing Microbial Communities, p. 165).

MICROBIAL CULTIVATION

The biorefinery environment is complex, and cultivation technologies must be capable of reproducing critical aspects of industrial systems. Efficient conversion of biomass to liquid fuels will require new and innovative approaches for controlling cultivation and simultaneously monitoring cell physiological states and metabolic processes under a range of conditions. Such capabilities are needed to identify gene regulatory and metabolic networks and to develop and evaluate modes of cell metabolism. Obtaining a systems-level understanding of fermentation organisms will require, in some cases, thousands of samples from single- and multiple-species cultures; and technologies for continuously monitoring and controlling culture conditions and interrogating the physiological state of microbial cells. Relatively homogeneous and complex microbial populations will be tested in physically and chemically heterogeneous environments such as those associated with solids (plant biomass). Necessary infrastructure will support cultivation at scales sufficient to obtain adequate amounts of sample for analysis and to grow microbial cells in monocultures and in nonstandard conditions (e.g., in association with solids or biofilms). These cultivation systems will be enhanced by advanced computational capabilities that allow simulation of cultivation scenarios and identification of critical experimental parameters.

Most industrial bioconversions rely on pure cultures while, in contrast, environmental bioconversions are more commonly catalyzed by mixed populations or communities with "specialists" working together in an apparently integrated and stable fashion. A fundamental question is, Are there stable, self-regulating multiplex solutions for biofuels? In this regard, new techniques are needed to understand and reproduce composition and function in stabilized mixed cultures (see sidebar, Laboratory Cultivation Techniques to Simulate Natural Community Structure, this page).

Biological systems are inherently inhomogeneous; measurements of the organism's average molecular-expression profile for a cell population cannot be related with certainty to the expression profile of any particular cell. For example, molecules found in small amounts in samples with mixed microbial species may be expressed either at low levels in most cells or at higher levels in only a small fraction of cells. Consequently, as a refinement, new analytical and imaging techniques will be necessary to interrogate individual cell physiological states in heterogeneous culture systems. For experimentation in all aspects of work with omic tools, production of high-quality reproducible samples is paramount.

LABORATORY CULTIVATION TECHNIQUES TO SIMULATE NATURAL COMMUNITY STRUCTURE

To identify the function of genes preferentially expressed by specific populations in a structured microbial community, such as those deconstructing biomass in soils or in a bioreactor, new cultivation techniques are being devised. During the past decade, researchers have developed reactors in which biofilms can be imaged using confocal scanning laser microscopy (CSLM) and other light-microscopic techniques (Wolfaardt et al. 1994). When combined with fluorescent in situ hybridization to distinguish populations of cells in multipopulation biofilms and fluorescent reporters (green fluorescent protein) of functional gene expression, CSLM has been used to demonstrate how gene expression by one population affects gene expression in another proximally located population (Moller et al. 1998).

The mobile pilot-plant fermentor shown here has a 90-L capacity and currently is used to generate large volumes of cells and cell products such as outer-membrane vesicles under highly controlled conditions. Future generations of fermentors will be more highly instrumented with sophisticated imaging and other analytical devices to analyze interactions among cells in microbial communities under an array of conditions.

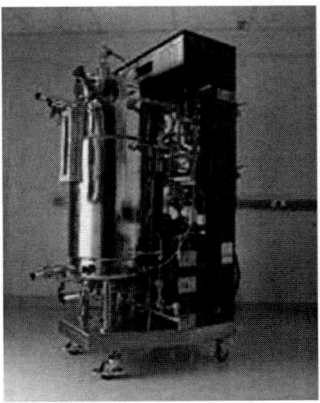

Pacific Northwest National Laboratory

DATA INFRASTRUCTURE

Progress toward efficient and economic processes for converting biomass to biofuels will benefit from a research-data network able to manage, preserve, query, and efficiently disseminate large amounts of experimental and analytical data, as outlined in the GTL

Roadmap. Because many research activities are data intensive and will engage a large number of investigators from national laboratories, academia, and the private sector, these activities will benefit from a distributed data environment.

Consisting of federated databases and repositories, the data environment will comprise descriptive, quantitative, and visual information, method libraries, query and data-mining tools, and a communication network. This environment provides controlled sharing and management and communication of biochemical, genetic, and other types of biological data and information. When available through a single portal, such a cyberinfrastucture can have a dramatic impact in efficient use of new knowledge. Additional benefits from efficient data sharing include fostering collaboration among projects, facilitating standards development, and accelerating implementation of new enabling technologies. GTL also will need a centralized research data network for model plants and specific biomass crop plants.

Data-infrastructure requirements of biomass-to-biofuels research align extensively with the Integrated Computational Environment for Biology described in the GTL Roadmap (see Figure 2. GTL Integrated Computational Environment for Biology, this page), reproduced as a programmatic subset of the GTL information infrastructure. Biomass-to-biofuels research will require multiple databases, imaging archives, and LIMS in each of its major research areas: Feedstock for Biofuels, Deconstructing Feedstocks to Sugars, and Sugar Fermentation to Ethanol.

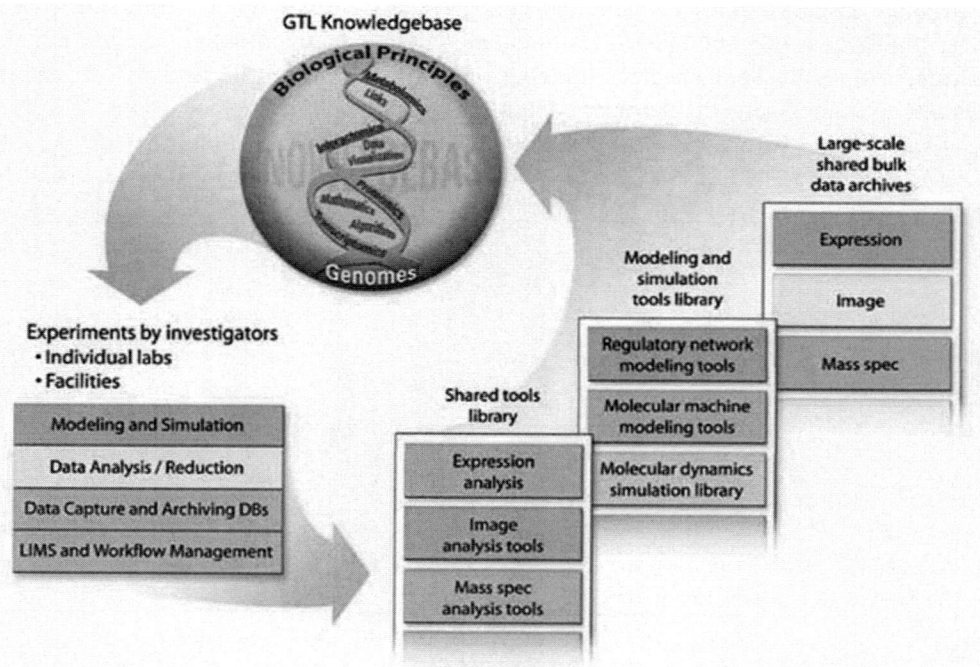

Source: GTL Roadmap, p. 83 (http://doegenomestolife.org/roadmap/).

Figure 2. GTL Integrated Computational Environment for Biology: Using and Experimentally Annotating GTL's Dynamic Knowledgebase.
At the heart of this infrastructure is a dynamic, comprehensive knowledge- base with DNA sequence code as its foundation. Offering scientists access to an array of resources, the knowledgebase will assimilate a vast range of microbial and plant data and knowledge as they are produced.

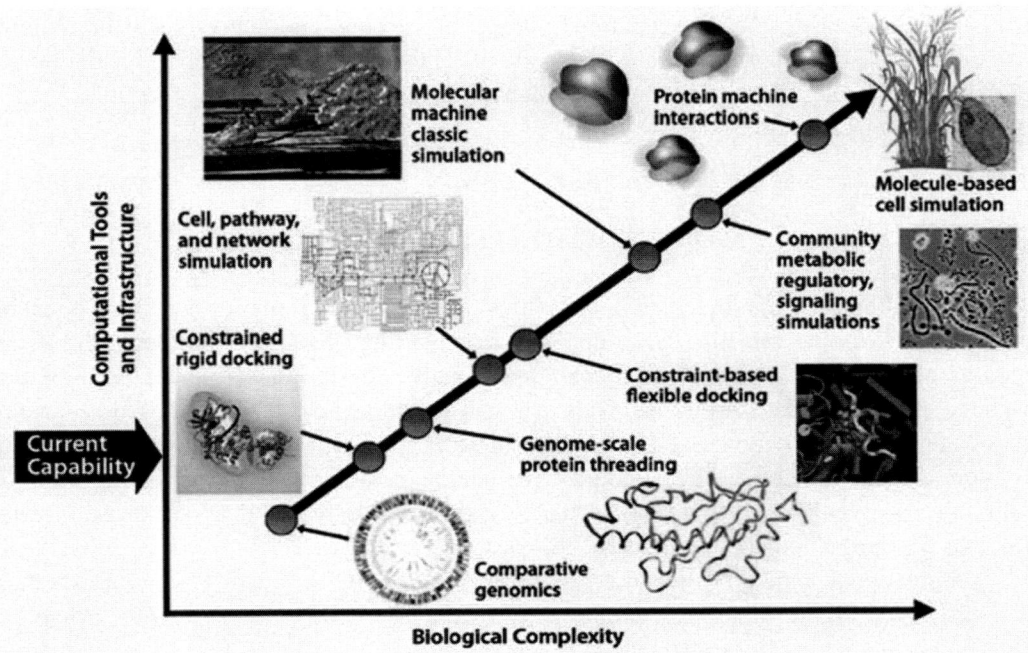

Source: GTL Roadmap, p. 89 (www.doegenomestolife.org/roadmap/).

Figure 3. From Genome Data to Full Cell Simulation. This concept diagram schematically illustrates a path from basic genome data to a more detailed understanding of complex molecular and cellular systems. New computational analysis, modeling, and simulation capabilities are needed to meet this goal. The points on the plot are very approximate, depending on the specifics of problem abstraction and computational representation. Research is under way to create mathematics, algorithms, and computer architectures for understanding each level of biological complexity.

For example, research on deconstructing feedstocks will necessitate (1) LIMS for tracking and documentation of samples for enzymatic pretreatment experiments and high-throughput compositional analysis of cell-wall material; enzyme databases for characterizing such key enzymes as glycohydrolases, esterases, and ligninases; metagenomic sequence databases for characterizing lignocellulolytic and other microbial communities; imaging repositories for a vast array of microscopies including optical, X ray, electron, and atomic force; and archives for molecular dynamic simulations of the molecular machinery that breaks down plant cell-wall components.

Feedstock and fermentation research goals have comparable data-infrastructure needs, with descriptive and quantitative data types differing mainly in experimental and analytical methods employed. Proteomic, metabolomic, transcriptomic, and plant and microbial genomic data will require corresponding databases and integration within the GTL Knowledgebase.

A variety of computational tools will be needed to support biomassto-biofuels research and such enabling technologies as proteomics and imaging. For example, computational tools are necessary for streamlining the conversion of isotopomer flux-tracing data into flux distributions for subsequent analysis; the process currently is limited by the computational sophistication of required methods rather than by the ability to generate raw data. High-throughput, automated image acquisition, storage, processing, and analysis will be required for examining data on molecular machines in vivo. These tools also will aid in localizing and validating complexes, dynamics, docking, intercellular communication, extracellular matrix, and metabolite distribution.

COMPUTATIONAL MODELING

Computational technologies will be critical to the overall platform that will enable success in developing effective technologies for converting biomass to biofuels. Significant amounts of data generated at a number of investigative levels must be made accessible to investigators at different institutions. Data sets will need to be analyzed and used to develop and evaluate computer models that will contribute to guiding overall decision making and program design. These decisions could involve the best lignocellulosic composition of biomass for a given process and how to create it as well as designing enzymes or engineering microbes for optimal ethanol production.

Computational modeling technologies and tools will be needed to address these challenges (see Figure 3. From Genome Data to Full Cell Simulation, p. 169). Some key needs and examples are discussed below.

Modeling: Genome Sequence Analysis

A wide range of organisms across many different kingdoms of life— including plants, fungi, and bacteria—are relevant to biomass conversion to biofuels. With the power of modern genome sequencing and capabilities at DOE's Joint Genome Institute, obtaining genome sequences for essentially any organism is possible. Research will use sequence-analysis tools surpassing traditional genome annotation. The complexity of biomass conversion to biofuels will necessitate the evaluation and integration of sequence data from multiple organisms. The need to establish functional roles for a multitude of proteins involved in mediating plant cell-wall biology and defining metabolic and regulatory pathways will require robust bioinformatic capabilities. Within each genome sequence is a wealth of information that can be mined and accessed to improve our knowledge about these organisms. In some cases, traditional bioinformatic approaches may be sufficient. Many areas, however, still need research advances.

Sequence-based analysis methods and tools are vital in many areas. Creation of an improved-quality standardized pipeline for genome-sequence annotation, including definition of open reading frames and functional assignment of genes, is a critical need for essentially all sequenced organisms that are subjects of GTL mission-directed research. These annotations need to be updated continuously because advancements can be rapid in many areas and should be applied to single organisms, including plants, and metagenomes of soil ecosystems and fermentative consortia. Specific to biomass-to-biofuels research are computational methods for improving analysis of metagenomic data to identify new enzymes related to lignocellulose degradation and fermentation. Directed evolution has the potential to rapidly evolve new and useful microbial strains for fermentation reactions. Comparative genomics of evolved microbes following resequencing will identify specific changes in the genome. Comparative analyses of multiple plant genomes will generate family clusters of gene components mediating cell-wall biosynthesis (i.e., enzymes, regulatory factors, and processing proteins). A more generic requirement is enhanced annotation related to recognition of interacting proteins, regulatory signals, and structures.

Modeling: Molecular

A number of critical biochemical processes, occurring at the molecular scale, govern the overall conversion of biomass to biofuels. These processes include, for example, mobilization of building blocks for plant cell-wall synthesis, machinery for cell-wall formation in plants and their biophysical characteristics, degradation of lignocellulosic materials by enzymes, and transport of sugars into microbes for further metabolism and conversion to desired end products. In combination with structural information, mechanistic models of plant-derived polymeric substrates and their enzymatic conversion are needed to identify "bottlenecks" in bioconversion of lignocellulosic biomass. Gaining a detailed understanding of these mechanisms will help in designing superior technologies to address key barriers. As a component of this research, new computer programs and codes are needed for ultralarge biological models of more than a million atoms to enable the analysis of complex molecular machines such as the cellulosome.

Molecular-scale models are critical to a number of biomass-to-biofuels aspects. For example, biophysical models of plant cell-wall composition are needed to delineate pathways for synthesis and transport of key cell- wall components. Our knowledge of fundamental polysaccharide-water interactions within lignin carbohydrate complexes will benefit greatly from molecular dynamics simulations, as will our understanding of water and chemical transport along or through the plant cell wall. Dynamic molecular models will aid in determining cell-wall enzymatic degradation and interactions among these enzymes and their substrates. Specifically, molecular models of such cell-wall–degrading enzymes as hemicellulases and ligninases will help to establish structure-function relationships for substrate interactions and guide the development of novel enzymes. Similarly, dynamic modeling of large molecular machines and their substrates on nano- to millisecond measures will provide insights into the structure and function of key protein complexes such as the cellulosome. Molecular modeling at a scale exceeding that for individual enzymes will enable understanding of critical events occurring at the interface between cellulose and an adhered cell. On the microbial side, models are needed to predict membrane composition and changes in response to stress induced by high concentrations of fermentation end products.

Modeling: Pathways and Networks

Molecular machines that carry out individual transformations and processes important in biomass-to-biofuels conversions often operate within much larger systems of interacting proteins and metabolic pathways in cells. Understanding, controlling, and manipulating the overall phenotypical state of these systems in either plant or microbial cells will be necessary to overcome identified technical barriers. Computational models of these cellular systems and pathways will facilitate integrative analysis of experimental omic data sets while also providing the basis for predictive simulations that can generate testable hypotheses. To increase the efficiency of fermentation reactions, improved metabolic and regulatory models are needed to understand mechanisms that control glycolytic flux and its impact on cellular metabolism. Because careful alterations of growth, energy, and redox conditions often are required to optimize fermentation reactions, robust models focused on analysis and regulation of cellular energetics are necessary. Regulatory and metabolic modeling at community and individual levels is essential for identifying and understanding gene regulatory networks.

A number of specific biomass-to-biofuels topics will require systems-level models and simulation methods. Genome-scale models of key industrial microbes are needed to understand metabolism details and allow rational design for improved biofuel production, including ways to identify novel biotransformation routes. Dynamic pathway models incorporating key enzyme kinetics for ethanol-producing pathways and other relevant cellular subsystems are a specific need. Methods also are required to assess physicochemical limitations of cellular systems and enzyme components to determine maximum-achievable metabolism rates (e.g., identify rate- limiting steps, both kinetic and diffusion limited). From the plant perspective, detailed pathway models of cell-wall biosynthesis in the context of overall plant metabolism are needed to generate testable hypotheses for controlling cell-wall composition. These models should incorporate information about catalytic activity, gene expression, mechanisms of control, and interspecies variations.

Generic modeling capabilities, many of which are required in the broader GTL program, include integrated modeling of metabolic pathways and regulatory networks and simulation of their functional capabilities. In support, automated techniques are needed to integrate data sets and rapidly reconstruct metabolic pathways and gene regulatory networks. Methods also will be required to incorporate the next level of complexity into systems-level cell models by adding information from cellular-component imaging. Other methods will account for the impact of protein and enzyme spatial localization within integrated models of cellular systems.

Modeling: Biorefinery Process

The long-term vision for biomass-to-biofuels research involves all cellular- and molecular-based procedures within an integrated biorefinery. To determine design requirements and alternatives, the overall process should be modeled computationally. Modeling is important, as is obtaining experimental validation of large-scale designs that will enable implementation of genomic and fundamental science research. In addition, physical properties and material handling including feed, mass, and fluidic transport parameters need to be determined. Without these parameters, designing bioprocessing facilities—including bioreactors, heat exchange, filtration, bioseparations, centrifugation, pumps, valves, and other chemical- engineering unit operations—is difficult and not comprehensive. This capability encompasses economic models and those that assess the net carbon balance impact of different processing schemes and the advantages and disadvantages of various coproduct processes. Models do exist for some of these processes, but they are not readily available to the community. To allow researchers to assess the feasibility of different design concepts, results and methodologies should be readily accessible and subject to continuous feedback-based improvements.

Process modeling will be very useful in evaluating the feasibility of consolidated bioprocessing. Considering the goal to provide the science underpinning one or more commercial processes for producing liquid fuels from biomass, a proper evaluation of most-probable future scenarios is needed. For example, one scenario envisioned for more efficiency and economy is simultaneous saccharification and fermentation (SSF). Saccharification enzyme mixes currently are optimal at 50°C and at a pH between 4 and 6, so as indicated by other study groups, *"none of the current ethanol-producing microorganisms is suitable."* Process modeling will evaluate the importance and sensitivity of SSF vs a two-step process. Modeling is needed to demonstrate rational design of integrated biomass processing by predicting and then verifying overall hydrolysis yields for native and modified biomass species using different pretreatment chemistries, temperatures, and specific enzymes.

In summary, research is needed to develop process models for assessing mass balances and economic models for bioprocess engineering and to disseminate them for community benefit (see chapter, Bioprocess Systems Engineering and Economic Analysis, p. 181).

CAPABILITY SUITES FOR BIOENERGY RESEARCH AND FACILITY INFRASTRUCTURE

To address mission science needs in energy and environment, the DOE Office of Science has proposed establishing vertically integrated research centers that will draw upon the range of advanced, high-throughput technologies and information-management computing described in the GTL Roadmap. The first of these centers will focus on bioenergy. This section provides an overview of existing and developing GTL capabilities that will be required to accelerate biomass-to-biofuels research. The GTL program has established pilots of most key technologies required to perform systems biology that might be incorporated into research centers. Biological capabilities need to be investigated at many scales (see Figure 9. Understanding Biological Capabilities at All Scales Needed to Support Systems

Biology Investigations of Cellulosic Biomass, p. 21). Over the past decade, genomic sequencing has been established as a highly efficient production component at the DOE Joint Genome Institute. Such capabilities provide the genomic foundation for research described in this report.

DOE Joint Genome Institute

DOE JGI's mission is to provide integrated high-throughput sequencing and computational analyses to enable genomic-scale, systems-based scientific approaches to DOE challenges in energy and environment. JGI capabilities will have an immediate impact on biomass-to-biofuels science by providing high-quality genome sequences of relevant crop plants and microorganisms, including naturally occurring and engineered microorganisms used for biomass conversion and soil microbial communities as they relate to soil quality and sustainability.

Identification of genes that control cell-wall composition in biomass species and development of tools such as gene chips depend on the availability of genome sequences for biomass crop species. Since micro-RNAs are expected to play a role in controlling expression in many relevant genes, DNA sequencing must be comprehensive enough to identify all microRNAs in these species.

Implementating effective biomass conversion will require JGI to rapidly and cost-effectively determine the sequences of a number of organisms, consortia, and metagenomes. In addition to biomass crop plants, JGI could provide genome sequence for new species of white rot fungi and brown rot fungi, actinomycetes, and such other biofuel-relevant organisms as natural consortia and communities involved in biomass decomposition. Full metagenomic sequencing will elucidate hemicellulose and lignin breakdown observed in mixed microbial populations and will allow harnessing of their collective processes.

Production and Characterization of Proteins and Molecular Tags

Functional Capabilities

These capabilities would enable the high-throughput expression of proteins mediating cell-wall biosynthesis (laccases, peroxidases, and glycosyl transferases) in *Arabidopsis* and other model and feedstock plants. Most enzymes of interest in cell-wall biosynthesis, such as glycosyl transferases, are thought to be membrane associated and, therefore, difficult to purify and characterize by conventional methods. Heterologously expressed proteins would be characterized biochemically and used as a resource for first-pass identification of function and the generation of antibodies and tags. To improve enzymes for lignocellulose deconstruction, the protein production capability would be a resource for native and recombinant forms of enzymes and modified enzymes from directed evolution and from rational-design, site-directed mutagenesis approaches. For pilot and validation studies, significant quantities of protein would be produced. Proteins, natural and modified, could be used to rapidly reconstitute machines such as cellulases and cellulosomes for improved performance. Antibody resources would have multiple applications, including pinpointing cellular localization of target proteins; protein complex identification; quantification of natural functional diversity; and quantitation of molecular outcomes from breeding experiments. In addition, this resource would generate tags and probes for imaging cell-wall polymers and progress of enzymes that degrade these polymers.

Assays for many cell-wall synthesis and deconstruction enzymes will require new approaches to screen for activity. In particular, they will involve access to different polysaccharide substrates and capabilities. In combination with proteomic capabilities, the resource would provide valuable information about proteins and mechanisms through which biomass is converted to biofuels. This knowledge will enable engineered-organism combinations to optimize biofuel production.

These capabilities will be critical for generating and characterizing new proteins and biomarkers for imaging. For example, a collection of fluorescently labeled proteins for such specific model plants as *Arabidopsis* or *Brachypodium* would be particularly valuable.

Instrumentation and Methods

From genome sequences, protein production and characterization methods will express proteins and generate reagents for interrogating cell function. Specifically, the goal will be to create capabilities to produce on demand all proteins encoded in any genome; create molecular tags that allow each protein to be identified, located, and manipulated in living cells; gain insights into function; and perform biophysical and biochemical protein characterizations. Using high-throughput in vitro and in vivo techniques (i.e., cellular and cell free) will lower protein-production costs to levels that will allow comprehensive analysis of cellular proteins. These methodologies will be applied to the production and characterization of modified and native proteins. Products and analysis capabilities will be made available to scientists and technologists.

"Affinity reagents" will be generated in parallel with protein production, using many of the same technologies. Tagged proteins or nucleic acids will permit detection and tracking of individual proteins in living systems, including complex molecular assemblages; intracellular position of all proteins and their spatial dynamics; if secreted, extracellular localization and interaction with other community members; and techniques for manipulating protein activity in the environment.

Potential core instrumentation:

- Gene-synthesis and manipulation techniques.
- High-throughput microtechnologies for protein-production screening.
- Robotic systems for protein and affinity-reagent production and characterization.
- Computing for data capture and management, genomic comparative analyses, control of high-throughput systems and robotics, and production-strategy determination.

Characterization and Imaging of Molecular Machines

Functional Capabilities

These capabilities will enable identification and characterization of protein complexes responsible for essential functions of energy-relevant phenotypes, including cell-wall biosynthesis. Understanding the function of these cell-wall synthesizing complexes and their interactions with metabolic pathways that produce sugar nucleotides in the cytosol will be important for understanding polysaccharide biosynthesis. Identifying key protein- protein interactions (interactomics) involving such biomass-deconstruction enzymes as polysaccharidases, hemicellulases, ligninases, and esterases, along with interactions with their substrates, also is a priority. Because polysaccharides probably are synthesized by a multi-enzyme complex in the Golgi, that also will be identified and analyzed.

Advanced capabilities are needed to understand the structure and function of cellulase and cellulosome molecular machines and improve designer cellulosomes by linking essential enzymes to the desired substrate and eliminating superfluous proteins and reactions. These advanced capabilities include analysis and computation for probing interactions among cells, their associated secreted enzyme complexes, and cellulose; and particularly for understanding the biology occurring at the cell surface–cellulose interface.

Instrumentation and Methods

Capabilities will identify and characterize molecular assemblies and interaction networks (see Figure 4. Visualizing Interaction Networks, this page). Resources will isolate and analyze molecular machines from microbial cells and plants; image and localize molecular machines in cells; and generate dynamic models and simulations of the structure, function, assembly, and disassembly of these complexes. High-throughput imaging and characterization technologies will identify molecular machine components, characterize their interactions, validate their occurrence, determine their locations within the cell, and allow researchers to analyze thousands of molecular machines that perform essential functions inside a cell. The capability for completely understanding individual molecular machines will be key in determining how cellular molecular processes work on a whole- systems basis, how each machine is assembled in 3D, and how it is positioned in the cell with respect to other cellular components.

Potential core instrumentation:

- Robotic culturing technologies to induce target molecular machines in microbial systems and support robotic techniques for molecular-complex isolation.
- State-of-the-art mass spectroscopy and other techniques specially configured for identification and characterization of protein complexes.
- Various advanced microscopies for intracomplex imaging and structure determination.
- Imaging techniques for both spatial and temporal intra- and intercellular localization of molecular complexes.
- Computing and information systems for modeling and simulation of molecular interactions that lead to complex structure and function.

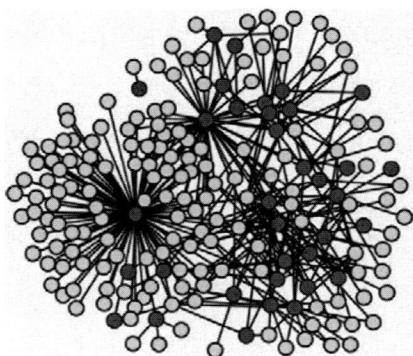

Source: GTL Center for Molecular and Cellular Systems at Oak Ridge National Laboratory and Pacific Northwest National Laboratory; previously published in GTL Roadmap, p. 68 (doegenomestolife.org/roadmap/).

Figure 4. Visualizing Interaction Networks.
Graphical maps display protein interaction data in an accessible form. These visualizations summarize data from multiple experiments and also allow quick determinations of proteins that might be core constituents of a particular protein complex and those that might play roles in bridging interactions among different complexes. The figure above, generated using Cytoscape, summarizes protein interactions in complexes isolated by affinity approaches from *Shewanella oneidensis*. Nodes (yellow or red circles) represent proteins.

Analysis of Genome Expression: The Omics

Functional Capabilities

An organism selectively produces portions of its proteome in response to specific environmental or intracellular cues. Studying its constantly changing protein expression thus leads to better understanding of how and why an organism turns genome portions "on" and "off." Identifying, quantifying, and measuring changes in global collections of proteins, RNA, metabolites, and other biologically significant molecules mediated by proteins—including lipids, carbohydrates, and enzyme cofactors—are important to this understanding. The ability to measure time dependence of RNA, protein, and metabolite concentrations will reveal the causal link between genome sequence and cellular function (see Figure 5. GeneProtein-Metabolite Time Relationships, this page). High-throughput omic tools (transcriptomics, proteomics, interactomics, metabolomics, glycomics, lignomics, and fluxomics) will be critical for systems-level investigation of plant cell-wall construction and processes whereby the cell wall is deconstructed and converted to ethanol. Multiple analytical capabilities will be seamlessly integrated with modeling to achieve the needed level of understanding.

Use of proteomics will allow comprehensive analysis of plant protein complexes, factors controlling their creation and function, and ensuing processes, beginning with the model plant *Arabidopsis*. The thousand or more proteins involved in cell-wall synthesis or modification are highly interacting or located in complexes. Knowing which proteins are in complexes before attempting to develop surrogate expression systems for enzyme characterization will be essential. Many such enzymes are membrane associated, so innovative methods for characterizing membrane complexes must be developed. These capabilities are vital to documenting the molecular makeup of living cell types in vascular and other tissues, including ray parenchyma and phloem.

MS and NMR would be used to develop very sensitive, high-throughput assays for glycosyl transferase activity. Synthesis, purification, and characterization of glycans needed as acceptors would be important.

A "glycochip" or related method for assaying enzymes involved in polysaccharide synthesis and modification also would be needed. High-throughput MS capabilities would be well suited for developing the analytical aspects of a glycochip. These resources also would benefit analysis of microbial systems used for plant cell-wall deconstruction and sugar fermentation. For example, such capabilities could quantify response of white rot fungi to changing culture conditions and provide a systems-level understanding for improving the energetics and carbon-allocation efficiency for cellulase and cellulosome production.

Optimizing microbial cultures for bioconversion will require tools that allow quantitative cellular characterization at the systems level, including those for global transcript, protein, and metabolite profiling in conjunction with metabolic and regulatory modeling. HTP methods to identify binding sites of global regulatory proteins will be required for models of global gene regulation. Additional HTP tools will monitor key metabolites that define cell redox and energy states [e.g., ATP, GTP, NAD(P)H, and NAD(P)] for fermentation optimization.

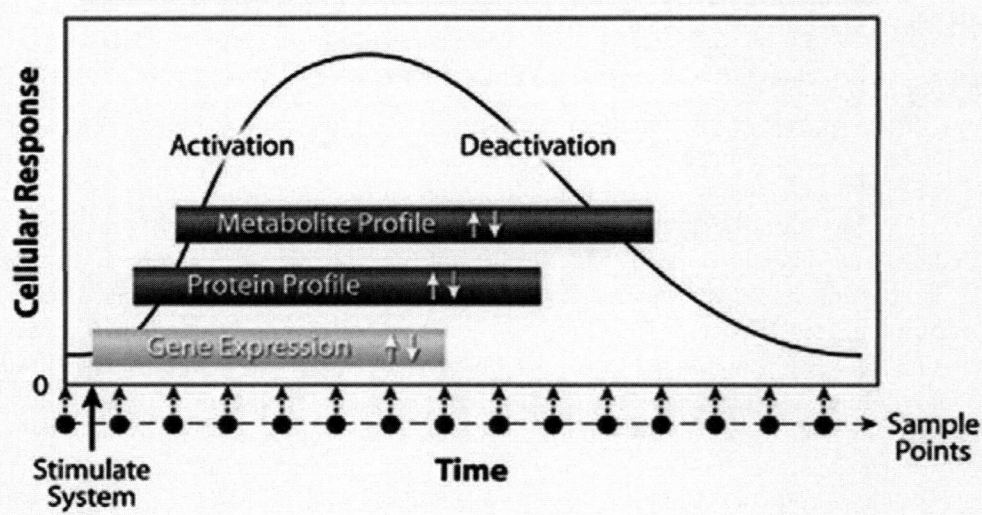

Source: Figure adapted from J. Nicholson et al., "Metabonomics: A Platform for Studying Drug Toxicity and Gene Function," *Nat. Rev.* 1, 153-61 (2002); previously published in GTL Roadmap, p. 158 (doegenomestolife.org/roadmap).

Figure 5. Gene-Protein-Metabolite Time Relationships.
To accurately establish causality among measured gene, protein, and metabolite events, sampling strategies must cover the full characteristic time scales of all three variables. Little is known about the time scale of gene, protein, and metabolite responses to specific biological stimuli or how response durations vary in genes and species.

Instrumentation and Methods

Consolidated omics will be capable of gaining insight into microbial functions by examining samples to identify (1) all proteins and other molecules created by a plant or individual or community of microbes under controlled conditions and (2) key pathways and other processes. Integrating these diverse global data sets, computational models will predict microbial functions and responses and infer the nature and makeup of metabolic and regulatory processes and structures.

Potential core instrumentation:

- Large farms of chemostats to prepare samples from highly monitored and controlled microbial systems under a wide variety of conditions.
- Numerous specialized mass and NMR spectrometers and other instruments capable of analyzing the molecular makeup of ensemble samples from thousands of diverse molecular species.
- High-performance computing and information capabilities for modeling and simulating plant or microbial-system functions under different scenarios. These studies will inform the design of systems-level experiments and help to infer molecular processes from ensuing data.

Analysis and Modeling of Cellular Systems

Functional Capabilities

Biomass complexity, coupled with intricate processes involved in conversion to ethanol, requires experimental capabilities and models for analyzing the biomass-enzyme-microbe system, specifically cell-wall synthesis and microbial decomposition of biomass. These capabilities also can contribute to determining the response of complex soil microbial communities to various cropping regimes.

On the basis of similarity with genes of known function, advanced computation will be used for comprehensive analysis of genes from model and biomass crop plants implicated in polysaccharide synthesis or modification. Acquisition and analysis of HTP gene-expression data from model and crop plants would be powerful in assigning probable function to genes implicated in cell-wall synthesis. As enzyme function emerges, a systems- level plant-cell model would incorporate biophysical and structural properties and knowledge about pertinent proteins. Models of this type would greatly facilitate the rational development of feedstock species based on "design principles," in which wall chemical composition is optimized as feedstock while plant productivity is simultaneously maximized.

These resources should include capabilities for generating models that enable improved cellulase production from near-term hosts such as *Trichoderma reesei* and proceed to bacterial and yeast systems that will require longer times for research and development. Other goals are to improve production of specific preferred components and foster a more global understanding of hyperproduction at industrial scales in industrial strains.

An integrated model of fermentative microorganisms and the ability to test it experimentally will support new concepts to produce liquid fuels and understand the producers' underlying metabolism and responses to stress in producing biofuels such as ethanol. Models and new rounds of modification and experimentation then will be used to alleviate stress responses and increase carbohydrate degradation, minimize cell growth and side products of synthesis, and maximize fuel production.

Metabolic flux analysis (as described earlier) is essential in improving understanding and manipulation of cellular metabolism. Possible analytical needs include appropriate NMR and MS instrumentation; stable isotopic labeling for key molecular moieties involved in metabolic processes; and synthesis of process intermediates. Intensive mathematical and computational power is required to achieve the final goal of flux estimation.

Instrumentation and Methods

Technologies and methodologies for analyzing cellular systems will be the capstone for ultimate analytical capabilities and knowledge synthesis to enable a predictive understanding of cell, organism, tissue, and community function critical for systems biology. Imaging methods will monitor proteins, machines, and other molecules spatially and temporally as they perform their critical functions in living cells and communities. Within their structures, microbial communities contain numerous microniches that elicit unique phenotypical and physiological responses from individual microbial species. The ability to analyze these niches and the microbial inhabitants within is crucial to our ultimate goal. This grand biological challenge must be addressed before scientists can predict the behavior of microbes and take advantage of their functional capabilities. Modeling resources will enable description of biological interactions with the physicochemical environment and predict how the system will evolve structurally and functionally.

Potential core instrumentation:

- Highly instrumented cultivation technologies to prepare structured microbial communities for simulating natural conditions in a highly controlled environment.
- Instruments integrating numerous analytical-imaging techniques to spatially and temporally determine, in a nondestructive way, relevant molecular makeup and dynamics of the community environment, community, and microbes that comprise it.
- Computing and information capabilities to model and simulate complex microbial systems, design experiments, and incorporate data.

CITED REFERENCES

Moller, G. M., et al. (1998). "In Situ Gene Expression in Mixed-Culture Biofilms: Evidence of Metabolic Interactions Between Community Members," *Appl. Environ. Microbiol. 64*, 721–32.

U.S. DOE. (2005). *Genomics: GTL Roadmap: Systems Biology for Energy and Environment*, U.S. Department of Energy Office of Science (doegenomestolife.org/roadmap/).

U.S. DOE. (2002). *Report on the Imaging Workshop for the Genomes to Life Program*, April 16–18, 2002, Office of Science, U.S. DOE (doegenomestolife.org/technology/imaging/workshop2002).

Wolfaardt, G. M., et al. (1994). "Multicellular Organization in a Degradative Biofilm Community," *Appl. Environ. Microbiol. 60*, 434–46.

In: Biological Barriers to Cellulosic Ethanol
Editor: Ernest V. Burkheisser

ISBN: 978-1-60692-203-3
© 2010 Nova Science Publishers, Inc.

Chapter 8

BIOPROCESS SYSTEMS ENGINEERING AND ECONOMIC ANALYSIS

Unites States Department of Energy

Complete process modeling should be initiated to guide the scientific work described in this Roadmap, including systems engineering and economic analyses, to evaluate the most-probable scenarios; to coordinate advances and needed research across the feedstock, deconstruction, and fermentation domains; and to reduce risk as the development cycle proceeds. Results and methodologies should be made available to the community through the web and should be subject to continuous improvements based on community feedback.

Biomass-conversion literature has many examples of pretreatment, hydrolysis, and fermentation systems that are technically effective but have no real chance to be competitive economically. A common tendency among biotechnologists is to attempt processes that will achieve, for example, the highest yields, rates, and titers. Process modeling, however, very often reveals that different combinations of these parameters are needed to produce a commercially viable process. Disciplined systems engineering and economic analysis using mass balances and standard analytical methods can eliminate ineffective approaches rather quickly, narrowing the focus to the most-promising options. Current advanced saccharification enzyme systems have demonstrated what a concerted, focused program of enzyme development can achieve. These systems, however, were designed for a specific acid-based pretreatment and the biomass raw material of corn stover. Other combinations of pretreatments and biomass materials will need to be analyzed and subjected to process engineering and economic analyses as they mature.

RESEARCH GOALS

This work seeks to integrate improvements in the molecular biology of plant materials and those in pretreatment, enzyme hydrolysis, and fermentation while conducting rigorous bioprocess systems engineering (incorporating new processes into a model refinery) and economic analysis. The approach integrates all elements of enhanced and cost-effective means to convert biomass into ethanol and other bioproducts.

Practical options for lignocellulose conversion to liquid fuels are sharply constrained by the need to produce fermentable sugar costing a few cents per pound, preferably with minimal sugar loss and degradation. Furthermore, in a biorefinery, all downstream processes will be affected by pretreatment choice. Process integration, therefore, is crucial. Finally, no evidence suggests that biological methods alone (i.e., without a thermochemical pretreatment step) will be able to obtain the high-yield, low-cost conversions of cellulose and hemicellulose required for economic viability. Research elements would include (1) correlation of cell-wall chemistry and structure with the efficiency of cellulase polysaccharide depolymerization, with and without pretreatment; (2) multidisciplinary studies of specific targeted lignocellulosic materials with promising pretreatments and advanced enzymes, using standard analytical methods and detailed mass balances; (3) examination of the effects of individual and combined enzymatic activities on pretreated substrates, with the primary technical-evaluation criterion of maximum sugar yield at minimum enzyme loadings; and (4) process engineering and economic modeling of these integrated pretreatment and hydrolysis studies to identify promising combinations and eliminate less-attractive options. This work must be coordinated with all domains of the research portfolio to have the desired effect—optimal process development.

Considering the goal to develop one or more commercial processes for ethanol production from biomass, a proper evaluation of most-probable future scenarios is needed. For example, several groups envision simultaneous saccharification and fermentation (SSF) for a more viable process. Since saccharification enzyme mixes work better at 50°C and a pH between 4 and 6, then as indicated by other study groups, "*none of the current ethanol producers' microorganisms is suitable.*" Under this scenario the parameters of saccharification must be included in designing new production microorganisms. Process modeling is needed to evaluate the importance and sensitivity of SSF vs a two-step process. Similarly, ethanol fermentations will be evaluated by three main parameters: Yield from carbohydrates, rate of production, and final titer.

Modeling will be very useful in evaluating the feasibility of consolidated bioprocessing. As an example, current specific-activity values of the three enzymes used today to degrade cellulose (i.e., exoglucanases, endoglucanases, and beta-glucosidases) may possibly allow prediction of the protein amount that a "consolidated strain" will have to produce per unit of time to degrade cellulose and liberate sugars quickly enough to sustain fast rates of ethanol production.

Milestones

Within 5 years or less

Initial research should address correlation of changes in plant cell-wall structure with ease of hydrolysis. This could be done through chemical composition or nanoscale and molecular-scale imaging (such as would be available from National Renewable Energy Laboratory). The relationship between changes in plant cell-wall structure (e.g., lignin composition) would enable better formulation of more-effective and more-productive advanced enzymes in their interactions with a solid substrate. Emphasis would be on coupling changes in plant cell-wall structure, changes in pretreatment conditions, and decreases in enzyme usage to achieve conversion extent and rate currently attained with much-higher enzyme loadings.

Within 10 years

Once scientists determine the impact of changes in plant cell-wall structure on hydrolysis extent and the ease with which a plant cell wall can be deconstructed, further mechanistic studies for improving the efficiency of enzyme hydrolysis will be carried out. Certain types of plant cell walls may have lignin content higher than or different from current biomass materials; utilization of enzymes including lignases, hemicellulases, and cellulases could be considered. A model would be helpful, for example, in which a lignin-rich material is converted efficiently into its fractions of lignin, celloligosaccharides, and xylans. Lignin could be used as a source of energy for a cellulose-to-ethanol processing plant. The proper composition balance in lignin to polysaccharides could make such a facility self-sufficient with respect to energy usage. Systems integration of these various steps will be carried out in a longer-term project. Scientists expect initial improvements in the efficiency of cellulose-to-ethanol conversion in an existing plant (such as a dry mill) to be possible within 10 years.

Within 15 years

Engineering, molecular evolution of new enzymes, and genetic modification of agronomic traits should be combined to yield a vertically integrated technology for converting biomass materials to ethanol and other bioproducts. Combining molecular biology with process engineering may yield plants that are robust in the field but easily accessible for conversion using specialized enzymes added after harvest. Transportation of biomass materials is known to be a key cost factor, so enzymes integrated into plant material could make it more compressible after harvest. Lignin-degrading enzymes may make possible high-lignin plants (including certain types of grasses) that are easily deconstructed once exposed to proper conditions in a processing plant, yielding an adequate amount for powering the facility.

The Role of GTL and OBP Facilities and Capabilities

The challenge in using advanced analytical and modeling capabilities is to integrate research vertically across several domains into a viable, technically and economically sound biorefinery concept. Research results must be rapidly assimilated into an overall systems model that will provide insight into the impact of new phenomena or processes; concomitant technologies must be incorporated into the system to optimize these effects. Such analysis will guide a vertically integrated research portfolio that supports all aspects of bioprocess engineering. A summary follows.

Protein Production

Protein production capabilities will be critical in generating and characterizing new proteins and biomarkers for imaging and for identifying key protein-protein interactions, including enzymes for lignin, polysaccharide, and hemicellulose ester degradation.

Proteomics

Proteomic measurements can provide valuable information about proteins and mechanisms by which monosaccharide conversion to ethanol and other products is achieved. Combinations of native and genetically modified organisms can achieve better conversion in existing ethanol facilities.

Cellular Systems

Cellular systems analysis and modeling may be employed to generate and monitor microbial cells and communities that would enable expression of new types of enzymes or proteins so sufficient quantities are available for research.

DOE Joint Genome Institute

DOE JGI will be important in discovering new microorganisms and enzymes that may be useful in converting more-recalcitrant forms of cellulosic materials to monosaccharides.

In: Biological Barriers to Cellulosic Ethanol
Editor: Ernest V. Burkheisser

ISBN: 978-1-60692-203-3
© 2010 Nova Science Publishers, Inc.

Chapter 9

APPENDIX A. PROVISIONS FOR BIOFUELS AND BIOBASED PRODUCTS IN THE ENERGY POLICY ACT OF 2005

United States Department of Energy

On July 29, 2005, Congress passed the Energy Policy Act of 2005,[1] and President George Bush signed it into law on August 8, 2005. The $14 billion national energy plan includes provisions that promote energy efficiency and conservation, modernize the domestic energy infrastructure, and provide incentives for both traditional energy sources and renewable alternatives. The following sections of the Energy Policy Act relate to biofuels and biobased products.

TITLE IX—RESEARCH AND DEVELOPMENT

Section 932. Bioenergy Program

Research, development, demonstration, and commercial application activities under this program will address biopower, biofuels, bioproducts, integrated biorefineries, crosscutting research and development in feedstocks; and economic analysis. Goals for DOE biofuel and bioproduct programs include partnering with industrial and academic institutions to develop: "(1) advanced biochemical and thermochemical conversion technologies capable of making fuels from lignocellulosic feedstocks that are price-competitive with gasoline or diesel in either internal combustion engines or fuel cell-powered vehicles; (2) advanced biotechnology processes capable of making biofuels and bioproducts, with emphasis on development of biorefinery technologies using enzyme-based processing systems; (3) advanced biotechnology processes capable of increasing energy production from lignocellulosic feedstocks, with emphasis on reducing the dependence of industry on fossil fuels in manufacturing facilities; and (4) other advanced processes that will enable the development of cost-effective bioproducts, including biofuels." Through Integrated Biorefinery Demonstration Projects,

DOE will demonstrate the commercial application of integrated biorefineries that use a wide variety of lignocellulosic feedstocks to produce liquid transportation fuels, high-value biobased chemicals, electricity, substitutes for petroleum-based feedstocks, and useful heat.

Section 941. Amendments to the Biomass Research and Development Act of 2000

This section presents several changes in the wording of the Biomass Research and Development Act of 2000. Some important changes include redefining the objectives of the Biomass Research and Development Initiative to state that the initiative will develop "(1) technologies and processes necessary for abundant commercial production of biobased fuels at prices competitive with fossil fuels; (2) high-value biobased products... to enhance the economic viability of biobased fuels and power... and [serve] as substitutes for petroleum-based feedstocks and products; and (3) a diversity of sustainable domestic sources of biomass for conversion to biobased fuels and biobased products." Another important amendment introduces four technical areas that the initiative will address in its research and development activities: (1) develop crops and systems that improve feedstock production and processing, (2) convert recalcitrant cellulosic biomass into intermediates that can be used to produce biobased fuels and products, (3) develop technologies that yield a wide range of biobased products that increase the feasibility of fuel production in a biorefinery, and (4) analyze biomass technologies for their impact on sustainability and environmental quality, security, and rural economic development.

Section 942. Production Incentives for Cellulosic Biofuels

This section calls for the establishment of a program that provides production incentives on a per-gallon basis for cellulosic biofuels. The purpose of the incentive program is to ensure that 1 billion gallons in annual cellulosic biofuel production are achieved by 2015; cellulosic biofuels are cost-competitive with gasoline and diesel; and small feedstock producers and rural small businesses are full participants in the cellulosic biofuel industry.

Section 977. Systems Biology Program

This section provides for the establishment of a research program in microbial and plant systems biology, protein science, and computational biology to support DOE energy, national security, and environmental missions. Funds will be available for projects to plan, construct, or operate special instrumentation or facilities for researchers in systems biology and proteomics and associated biological disciplines. Biomedical research on human cells or human subjects is prohibited.

TITLE XV—ETHANOL AND MOTOR FUELS

Section 1501. Renewable Content of Gasoline (Renewable Fuels Standard)

This section amends Section 211 of the Clean Air Act. An important amendment is the establishment of the renewable fuel program. In this program, gasoline sold in the United States is required to be mixed with increasing amounts of renewable fuel (usually ethanol) on an annual average basis. In 2006, 4 billion gallons of renewable fuels are to be mixed with gasoline, and this requirement increases annually to 7.5 billion gallons of renewable fuel by 2012. For 2013 and beyond, the required volume of renewable fuel will include a minimum of 250 million gallons of cellulosic ethanol.

Section 1505. Public Health and Environmental Impacts of Fuels and Fuel Additives

This section amends Section 211(b) of the Clean Air Act by requiring study of the impacts of increased use of fuel additives (e.g., ethanol and other chemicals) on public health, air quality, and water resources.

Section 1506. Analysis of Motor Vehicle Fuel Changes

This section amends Section 211 of the Clean Air Act by requiring study of the effects of ethanol-gasoline mixes on permeation of fuel molecules into plastic and rubber components of fuel systems and the evaporative emissions resulting from this permeation.

Section 1511. Renewable Fuel

This section amends the Clean Air Act by adding a section that allows funds to cover loan guarantees for no more than four projects commercially demonstrating the feasibility of producing cellulosic ethanol. Each project is expected to produce 30 million gallons of cellulosic ethanol annually. Funds also have been authorized for a resource center that further develops bioconversion technologies for ethanol production.

Section 1512. Conversion Assistance for Cellulosic Biomass, Waste-Derived Ethanol, Approved Renewable Fuels

This section amends Section 211 of the Clean Air Act by authorizing DOE to provide grants for building production facilities to producers of cellulosic ethanol and other approved renewable fuels.

Section 1514. Advanced Biofuel Technologies Program

This section authorizes the EPA Administrator to establish a program to demonstrate advanced technologies for the production of alternative transportation fuels. Priority is given to projects that increase the geographic distribution of alternative fuel production and make use of underutilized biomass feedstocks.

Section 1516. Sugar Ethanol Loan-Guarantee Program

This section authorizes DOE to issue loan guarantees to projects that demonstrate the feasibility of producing ethanol from sugar cane or sugar cane by-products.

TITLE II—RENEWABLE ENERGY

Section 208. Sugar Cane Ethanol Program

The Environmental Protection Agency is to establish a $36 million program to study sugar cane, production of ethanol from sugar cane, and sugar cane by-products. The project will be limited to sugar and ethanol producers in Florida, Louisiana, Texas, and Hawaii. Information will be gathered on how to scale up production once the sugar cane industry is ready to locate sites for and construct ethanol-production facilities.

Section 210. Grants to Improve the Commercial Value of Forest Biomass for Electric Energy, Useful Heat, Transportation Fuels, and Other Commercial Purposes

The Secretary of Agriculture or the Secretary of the Interior may provide grants to any person in a preferred community (e.g., Indian tribe, town or unit of local government with fewer than 50,000 individuals, or one that has been determined to pose a fire, disease, or insect-infestation hazard to federal or Indian land). Owners of facilities that use biomass to produce electricity, useful heat, or transportation fuels may receive grants of up to $20 per green ton of biomass delivered to offset the cost of raw material. Grants of up to $500,000 may be awarded to offset the costs of developing research opportunities that improve the use of or add value to biomass. For each fiscal year from 2006 through 2016, $50 million is authorized to be appropriated to carry out this section.

End Notes

[1] U.S. Congress. 2005. Energy Policy Act of 2005, Pub. L. 109- 58 (www.epa.gov/OUST/fedlaws/publ_109-058.pdf).

In: Biological Barriers to Cellulosic Ethanol
Editor: Ernest V. Burkheisser

ISBN: 978-1-60692-203-3
© 2010 Nova Science Publishers, Inc.

Chapter 10

APPENDIX B. WORKSHOP PARTICIPANTS

United States Department of Energy

Ackerman, Eric
Pacific Northwest National Laboratory
eric.ackerman@pnl.gov

Adams, Justin
BP PLC
justin.adams@uk.bp.com

Armstrong, Katherine
Dow Agrosciences LLC
karmstrong@dow.com

Atalla, Rajai
33/3 14 Forest Products Laboratory
rhatalla@wisc.edu

Baldwin, Sam
Office of Energy Efficiency and Renewable Energy
U.S. Department of Energy
sam.baldwin@ee.doe.gov

Bayer, Ed
Department of Biological Chemistry
Weizmann Institute of Science
ed.bayer@weizmann.ac.il

Bownas, Jennifer
Genome Management Information System
Oak Ridge National Laboratory

bownasjl@ornl.gov

Brady, John
Department of Food Science
Cornell University
jwb7@cornell.edu

Bull, Stan
National Renewable Energy Laboratory
stan_bull@nrel.gov

Cameron, Doug
Cargill, Incorporated
doug_cameron@cargill.com

Casey, Denise
Genome Management Information System
Oak Ridge National Laboratory
caseydk@ornl.gov

Cavalieri, Ralph
Washington State University
cavalieri@wsu.edu

Chapple, Clint
Department of Biochemistry
Purdue University
chapple@purdue.edu

Chum, Helena
National Renewable Energy Laboratory
helena_chum@nrel.gov

Cleary, Michael
San Diego Supercomputer Center
University of California, San Diego
mcleary@sdsc.edu

Collart, Frank
Argonne National Laboratory
fcollart@anl.gov

Colson, Steven
Fundamental Science Directorate
Pacific Northwest National Laboratory
steven.colson@pnl.gov

Cotta, Mike
National Center for Agricultural Utilization Research
Agricultural Research Service
U.S. Department of Agriculture
cottama@ncaur.usda.gov

Dale, Bruce
Department of Chemical Engineering and Materials Science
Michigan State University
bdale@egr.msu.edu

Davison, Brian
Life Sciences Division
Oak Ridge National Laboratory
davisonbh@ornl.gov

Dean, William
Danisco Genencor International
bdean@danisco.com

Donohue, Tim
Bacteriology Department
University of Wisconsin, Madison
tdonohue@bact.wisc.edu

Drell, Daniel
Biological and Environmental Research
Office of Science
U.S. Department of Energy
daniel.drell@science.doe.gov

Ferrell, John
Biomass Program
Office of Energy Efficiency and Renewable Energy
U.S. Department of Energy
john.ferrell@hq.doe.gov

Foust, Tom
National Renewable Energy Laboratory
thomas_foust@nrel.gov

Fredrickson, Jim
Pacific Northwest National Laboratory
jim.fredrickson@pnl.gov

Gonzalez, Ramon
Department of Chemical and Biomolecular Engineering
Rice University
ramon.gonzalez@rice.edu

Greene, Rich
Office of International Research Programs
Agricultural Research Service
U.S. Department of Agriculture
richard.greene@nps.ars.usda.gov

Hames, Bonnie
National Renewable Energy Laboratory
bonnie_hames@nrel.gov

Harrison, Maria
Boyce Thompson Institute for Plant Research
mjh78@cornell.edu

Heineken, Fred
National Science Foundation
fheineke@nsf.gov

Hennessey, Susan
DuPont Central Research and Development
susan.m.hennessey@usa.dupont.com

Himmel, Mike
National Bioenergy Center
National Renewable Energy Laboratory
mike_himmel@nrel.gov

Hladik, Maurice
Iogen Corporation
mauriceh@iogen.ca

Houghton, John
Biological and Environmental Research
Office of Science
U.S. Department of Energy
john.houghton@science.doe.gov

Ingram, Lonnie
Florida Center for Renewable Chemicals and Fuels
Department of Microbiology and Cell Science
University of Florida

ingram@ufl.edu

Jacobs-Young, Chavonda
National Research Initiative
Cooperative State Research, Education, and Extension Service
U.S. Department of Agriculture
cjacobs@csrees.usda.gov

Jofuku-Okamura, Diane
Division of Biological Infrastructure
National Science Foundation
dbipqr@nsf.gov

Kaempf, Doug
Biomass Program
Office of Energy Efficiency and Renewable Energy
U.S. Department of Energy
douglas.kaempf@ee.doe.gov

Kahn, Michael
Basic Energy Sciences
Office of Science
U.S. Department of Energy
michael.kahn@science.doe.gov

Kaleikau, Ed
National Research Initiative
Cooperative State Research, Education, and Extension Service
U.S. Department of Agriculture
ekaleikau@csrees.usda.gov

Keasling, Jay
Physical Biosciences Division
Lawrence Berkeley National Laboratory
jdkeasling@lbl.gov

Keegstra, Ken
MSU-DOE Plant Research Laboratory
Michigan State University
keegstra@msu.edu

Klembara, Melissa
Biomass Program
Office of Energy Efficiency and Renewable Energy
U.S. Department of Energy
melissa.klembara@hq.doe.gov

Knotek, Mike
m.knotek@verizon.net

Ladisch, Mike
Laboratory of Renewable Resources Energy
Purdue University
ladisch@purdue.edu

Lohman, Kent
Biological and Environmental Research
Office of Science
U.S. Department of Energy
kenton.lohman@science.doe.gov

Lynd, Lee
Thayer School of Engineering
Dartmouth College
lee.r.lynd@dartmouth.edu

Mansfield, Betty
Genome Management Information System
Oak Ridge National Laboratory
mansfieldbk@ornl.gov

Matteri, Bob
Agricultural Research Service
U.S. Department of Agriculture
rmatteri@pw.ars.usda.gov

McLean, Gail
National Research Initiative
Cooperative State Research, Education, and Extension Service
U.S. Department of Agriculture
gmclean@csrees.usda.gov

Michaels, George
Bioinformatics and Computational Biology
Pacific Northwest National Laboratory
george.michaels@pnl.gov

Miranda, Amy
Biomass Program
Office of Energy Efficiency and Renewable Energy
U.S. Department of Energy
amy.miranda@hq.doe.gov

Mitchinson, Colin
Danisco Genencor Intl. Inc.
cmitchinson@danisco.com

Moorer, Richard
Office of Energy Efficiency and Renewable Energy
U.S. Department of Energy
richard.moorer@hq.doe.gov

Morrison, Mark
Department of Microbiology
Ohio State University
morrison.234@osu.edu

Palmisano, Anna
Cooperative State Research, Education, and Extension Service
U.S. Department of Agriculture
anna.palmisano@usda.gov

Patrinos, Ari
Biological and Environmental Research
Office of Science
U.S. Department of Energy
ari.patrinos@science.doe.gov

Ragauskas, Art
Institute of Paper Science and Technology
Georgia Institute of Technology
art.ragauskas@ipst.gatech.edu

Ralph, John
U.S. Dairy Forage Research Center
Cooperative State Research, Education, and Extension Service
U.S. Department of Agriculture
Department of Forestry
University of Wisconsin, Madison
jralph@wisc.edu

Remington, Karin
J. Craig Venter Institute

Sarkanen, Simo
Kaufert Laboratory
Department of Bio-Based Products
University of Minnesota, Twin Cities
sarka001@umn.edu

Schilling, Christophe
Genomatica, Inc.
cschilling@genomatica.com

Shanklin, John
Biology Department
Brookhaven National Laboratory
shanklin@bnl.gov

Shoemaker, Sharon
Department of Food Science and Technology
University of California
spshoemaker@ucdavis.edu

Shoham, Yuval
Department of Biotechnology and Food Engineering
Technion-Israel Institute of Technology
yshoham@tx.technion.ac.il

Smith, Lloyd
Department of Chemistry
University of Wisconsin, Madison
smith@chem.wisc.edu

Somerville, Chris
Department of Plant Biology
Department of Biological Sciences
Carnegie Institution of Washington
Stanford University
crs@stanford.edu

Stephanopoulos, Greg
Department of Chemical Engineering
Massachusetts Institute of Technology
gregstep@mit.edu

Stevens, Walt
Chemical Sciences, Geosciences, and Biosciences
Division Basic Energy Sciences
Office of Science
U.S. Department of Energy
walter.stevens@science.doe.gov

Stone, Bruce
Department of Biochemistry
La Trobe University

b.stone@latrobe.edu.au

Stults, Ray
National Renewable Energy Laboratory
ray_stults@nrel.gov

Tabita, Bob
Department of Microbiology
Ohio State University
tabita.1@osu.edu

Thomas, Steve
Ceres, Inc.
sthomas@ceres-inc.com

Thomassen, David
Biological and Environmental Research
Office of Science
U.S. Department of Energy
david.thomassen@science.doe.gov

Tuskan, Gerald
Environmental Sciences Division
Oak Ridge National Laboratory
tuskanga@ornl.gov

Uberbacher, Ed
Life Sciences Division
Oak Ridge National Laboratory
ube@ornl.gov

Valle, Fernando
Danisco Genencor, Inc.
fvalle@Danisco.com

Vogel, John
Western Regional Research Center
jvogel@pw.usda.gov

Vogel, Kenneth
Agricultural Research Service
U.S. Department of Agriculture
Department of Agronomy and Horticulture
University of Nebraska, Lincoln
kpv@unlserve.unl.edu

Weatherwax, Sharlene
Biological and Environmental Research Office of Science
U.S. Department of Energy
sharlene.weatherwax@science.doe.gov

Wheeler, Nick
Molecular Tree Breeding Services, LLC
nickwheeler@scattercreek.com

Wilson, David
Department of Molecular Biology and Genetics
Cornell University
dbw3@cornell.edu

Chapter 11

APPENDIX C. WORKSHOP PARTICIPANT BIOSKETCHES

United States Department of Energy

ERIC ACKERMAN

Eric Ackerman is a molecular biologist at Pacific Northwest National Laboratory, having earned a doctorate in biophysics from the University of Chicago in 1979. He was a Helen Hay Whitney Fellow at the Laboratory of Molecular Biology in Cambridge, England, for 3 years and then a staff scientist at NIH National Institute of Diabetes and Digestive and Kidney Diseases until 1996. He has been involved in developmental biology projects with *Xenopus laevis*, the methods by which toxins kill cells, mechanisms of nucleases, biochemistry of nucleotide excision DNA repair, and an extremely sensitive and quantitative assay to measure radiation effects on multiple kinds of DNA repair using as few as 3000 cells and 0.1-ng DNA.

Most relevant for the Biomass to Biofuels workshop have been Ackerman's studies of novel mechanisms for immobilizing enzymes in functionalized nanoporous materials for enhanced activity and stability. Recently, he began implementing high- throughput, cell-free production of proteins and their characterizations in hopes that this approach might yield optimized, stable protein complexes that could contribute to energy generation. He also was involved in strategic planning for Genomics: GTL production facilities, particularly for the Facility for Production and Characterization of Proteins and Affinity Reagents.

JUSTIN ADAMS

Justin Adams joined BP PLC in 2003 and currently is director of long-term technology strategy in BP's Office of the Chief Scientist. In this role he helps build and shape the strategic agenda, coordinates all long-term activities across the company to ensure alignment and balance, and oversees specific programs managed by the central technology function.

Before joining BP, Adams was founder and CEO of High Power Lithium, a Swiss company developing next-generation battery materials for hybrid electric vehicles in collaboration with Toyota. He also was an advisor to Konarka Technologies, a Massachusetts-based startup developing next-generation solar cells using conducting polymers and nanostructured materials. He previously worked as a consultant with Arthur D. Little, ultimately leading its Advanced Energy Systems practice in Europe. Arthur D. Little delivers strategic and technoeconomic consulting on emerging energy technologies to many of the world's leading energy majors.

Adams holds joint honors in management and technology from the University of Bath (England) and the University of Richmond (Virginia).

KATHERINE ARMSTRONG

Katherine Armstrong is global leader for trait genetics and technologies R&D at Dow AgroSciences in Indianapolis, Indiana. She earned her bachelor's degree from the University of Virginia and master's in molecular and population genetics from the University of Georgia. She has been an R&D scientist with Dow for 25 years and has studied plant gene expression at the molecular and cellular levels. Currently she oversees the development of corn traits through product launch. She holds seven U.S. patents in the area of plant gene expression and has written numerous publications. Relevant areas of research include optimization of corn genetics for ethanol extraction from both grain and cellulosic feedstocks.

RAJAI H. ATALLA

Rajai Atalla received his bachelor's degree from Rensselaer Polytechnic Institute in 1955 and master's and doctorate in chemical engineering and physics from the University of Delaware by 1960; his work focused on spectroscopic studies of flames. During 8 years at Hercules Research Center, he studied phase transitions in semicrystalline polymers and evaluated anomalous spectra of many compounds. He was first to recognize that anomalous proton nuclear magnetic resonance (NMR) spectra of (alkyl phosphito) hydrides of cobalt and iron indicated the occurrence of fluxional molecules. He also developed the first theoretical model for photodegradation of inorganic pigment.

As professor of chemical physics and engineering at the Institute of Paper Chemistry, Atalla pioneered the application of Raman spectroscopy to studies of celluloses. Finding accepted crystal structures inconsistent with Raman spectra, he investigated the 13C solid-state NMR spectra of native celluloses with David VanderHart of the National Institute of Standards and Technology. They determined that all native celluloses are composites of two forms—I_α and I_β (1984). With Umesh Agarwal, using a Raman microprobe, Atalla developed the first direct evidence of lignin orientation in secondary walls (1984).

In 1989, as head of chemistry and pulping research at the U.S. Department of Agriculture (USDA) Forest Service and adjunct professor in chemical and biological engineering at the University of Wisconsin, Madison, Atalla led development of inorganic analogs of lignin peroxidases for use in liquid-effluent–free pulping and bleaching systems. The processes were feasible economically, but the industry's economic condition led to suspension of the program. Freed of administrative responsibilities in 1999 and elevated to senior and pioneering research scientist in 2005, he returned to studies of molecular architecture in plant cell walls, with emphasis on secondary walls and native celluloses.

Atalla has published more than 150 papers, edited a book on cellulose structures, and is a fellow of the International Academy of Wood Science and of the Technical Association of the Pulp and Paper Industry. He received the Anselme Payen Award of the American Chemical Society's Cellulose Division as well as multiple USDA awards, including the Forest Service Chief's Distinguished Scientist Award.

ED BAYER

Ed Bayer is a professor in the Department of Biological Chemistry at the Weizmann Institute of Science, Rehovot, Israel. He was awarded a bachelor's degree in liberal arts from the University of Michigan in 1969, a master's in biology from Wayne State University in 1971, and a doctorate in biophysics from the Weizmann Institute of Science in 1976. Since the early 1970s, he has been involved in developing the avidin-biotin system as a general tool in the biological sciences. He was first to use biotinylation procedures for antibodies and other proteins and carbohydrates as well as avidin-conjugation and complexation techniques. The work initially was published in the mid-1970s, and many of the procedures are still in routine use today. He received the Sarstedt Award for his contributions to the avidinbiotin system for biomedical analysis.

Together with Raphael Lamed, Bayer introduced the cellulosome concept in the early 1980s. In 1999, he was organizer and cochair of the first Gordon Research Conference on Cellulases and Cellulosomes, and he served as chairman of the same conference in 2001. In 1994, he proposed the use of "designer cellulosomes" for biomass degradation and waste management and as a general tool in the biological sciences. Since then, he has worked systematically toward the controlled construction of such artificial cellulosomes via self-assembly and has produced a growing repertoire of divergent cellulosomal components for this purpose.

During his career, he has collaborated with groups in the United States, Canada, Holland, Belgium, Germany, Great Britain, France, Spain, Finland, Denmark, Guatemala, and the Republic of Georgia, and he has authored more than 250 articles and reviews in both fields. He coedited Vol. 184 on avidin-biotin technology in the *Methods in Enzymology* series and since 1999 has served as editor of the review journal *Biotechnology Advances*. In 2002, he was elected a fellow of the American Academy of Microbiology. He continues his work in both the avidin-biotin and cellulosome fields, and his interests still focus on protein engineering, nanobiotechnology, and the structural and functional consequences of protein-protein and protein-ligand interactions.

JOHN BRADY

John Brady is a professor in the Department of Food Science at Cornell University. He received a bachelor's degree in chemistry from the University of North Carolina, Chapel Hill, in 1975 and a doctorate in chemistry from the State University of New York at Stony Brook in 1980. During much of his graduate study, he was a visiting staff member at Los Alamos National Laboratory in New Mexico. He received his postdoctoral training in chemistry at Harvard University, working with Martin Karplus.

His research primarily involves biopolymer dynamics and hydration and the relationships among structure, conformation, and function in biological systems. Specific examples include the solution behavior of biopolymers, factors that determine secondary and tertiary structure in polymers, enzymatic reaction mechanisms, rational drug design, effects of point mutations in proteins, and the possibility of engineering desirable modifications in the function of wild-type proteins. His work uses techniques of computational theoretical chemistry to model properties of biopolymers and solutions numerically. These techniques, often called molecular mechanics, include computer graphics–based molecular docking, energy minimization and conformational energy calculations, and molecular dynamics simulations.

A principal focus of Brady's research is on carbohydrate structure, dynamics, and hydration. He has contributed to advances in carbohydrate modeling, including the first molecular dynamics simulations on a sugar, the first relaxed conformational energy map for a disaccharide, the first free-energy simulations of sugar energy differences in solution, and the first potential of mean- force or conformational free-energy map for a disaccharide. As an outgrowth of his primary interests in carbohydrates, he is studying carbohydrate interactions with proteins. In a current project, Brady is using molecular mechanics simulations to study the catalytic mechanism and mode of substrate binding in various cellulases, including E2 from *Thermomonospora fusca*, in the hope of designing a more active enzyme that could be produced by site-directed mutagenesis.

DOUG CAMERON

Doug Cameron received a bachelor's degree in biomedical engineering in 1979 from Duke University and a doctorate in biochemical engineering from MIT in 1986. He serves as director of biotechnology for Cargill Research, with an adjunct professorship in the Department of Chemical and Biological Engineering at the University of Wisconsin, Madison (UWM). From 1986 to 1998, Cameron was a professor in the Department of Chemical Engineering and an affiliate in the Molecular Biology Program at UWM. In 1996 he was a guest professor in the Institute for Biotechnology at the Swiss Federal Institute of Technology in Zurich. From 1979 to 1981, he held the position of biochemical engineer at Advanced Harvesting Systems, a plant biotechnology company funded by International Harvester.

Cameron is a fellow of the American Institute of Medical and Biological Engineering and is on the editorial boards of *Metabolic Engineering* and *Biomacromolecules*. He serves on the Minnesota Governor's Bioscience Council and the board of directors of Minnesota Biotechnology Industry Organization. He is a member of the MIT Biological Engineering visiting committee and on the managing board of the newly formed Society for Biological Engineering. Cameron also is a consulting professor in the Department of Chemical Engineering at Stanford University.

CLINT CHAPPLE

Clint Chapple received his doctorate in chemistry from the University of Guelph in 1989. After doing postdoctoral work with Chris Somerville at Michigan State University, he joined the faculty of Purdue University in the Biochemistry Department. Chapple's research in the area of lignin biosynthesis and plant secondary metabolism, using *Arabidopsis* as a model system, has earned him the title of Purdue University Scholar and a fellowship in the American Academy for the Advancement of Science. Research by the Chapple group helped to change the traditional paradigm of the role of ferulic and sinapic acids in building plant cell walls. Rather than contributing to the production of lignin, the group found that these two acids serve as end products in an essential biochemical pathway for cell-wall construction.

The phenylpropanoid pathway gives rise to a wide array of soluble metabolites in plants. These compounds participate in many plant defense responses and absorb potentially damaging UVB radiation. The pathway also generates the monomers required for lignin biosynthesis—ferulic acid and sinapic acid. Lignin is integrated into the plant secondary cell wall, where it provides structural rigidity to plant tissues and enables tracheary elements to withstand the tension generated during transpiration.

Chapple received the 2001 Agricultural Researcher Award from the Purdue School of Agriculture for his patented work in engineering plants to store and stabilize plastic monomer precursors in vacuoles.

HELENA L. CHUM

Trained in physical and industrial chemistry, Helena Chum has worked in bioenergy and renewable energy since 1979 at the Solar Energy Research Institute, now National Renewable Energy Laboratory (NREL), and has led R&D branches, divisions, and centers at NREL since 1992. Her general research involves technology development for conversion of biomass and a variety of organic wastes into biofuels, chemicals, electricity, and high-value materials. She has coauthored a book, 85 peer-reviewed publications, and 150 meeting papers. She has presented 100 invited lectures worldwide and jointly holds 18 patents.

Her specific research areas are biomass chemical analyses, standards development, and rapid spectrometric analysis methodologies; biomass and urban and plastic residue conversion to chemicals and biofuels; thermochemical conversion to multiple products; biomass fractionation; electrochemistry applied to biomass and derived compounds; environmental technologies; thermally regenerative electrochemical systems and fuel cells; technology development and government-industry-academia partnerships in R&D; and analyses of U.S. governmental biomass and hydrogen programs and their impact on commercial tools and systems to support recommendations for future programs.

Chum is a fellow of the American Association for the Advancement of Science for integrating industry-academia-government research partnerships in biomass and biobased materials and also of the International Academy of Wood Science for demonstrated leadership of biomass analysis and standards activities worldwide. She received a certificate of appreciation from the U.S. DOE Assistant Secretary of the Office of Energy Efficiency and Renewable Energy (EERE) for contributions and leadership in departmental, presidential, and congressional environmental initiatives, including the National Environmental Technology Strategy, and for dedication to EERE programs (1995).

MIKE CLEARY

Joseph Michael Cleary is Executive Division Director of Sciences R&D at the San Diego Supercomputer Center (SDSC), University of California, San Diego. He received his bachelor's degree in biology from Stanford University in 1970 and his doctorate in molecular biology from the University of California, Los Angeles, in 1980. At SDSC, he directs groups that support the cyber infrastructure needs of researchers by producing data systems and computational tools to facilitate discoveries in the natural sciences. His responsibilities include the initiation of interdisciplinary research projects for life science programs with biomedical and biology researchers at university, government, and independent research institutions.

Before joining SDSC in 2003, Cleary held biotechnology research and management positions for over 20 years at Merck and Monsanto, where he worked in fermentation and microbial- strain development, with emphasis on biosynthetic pathways for producing commercially valuable bacterial polysaccharides. He is an adjunct professor of biology at San Diego State University and serves as a consultant to several biotechnology businesses on matters related to industrial microbiology.

FRANK COLLART

Frank Collart is manager of the Robotic Molecular Biology Facility in the Biosciences Division at Argonne National Laboratory and Cloning and Expression group leader of the Midwest Center for Structural Genomics (MCSG). He received his doctorate in medical sciences from the Medical College of Ohio and master's degree in chemistry from Bowling Green State University. He has used cultured cell models to delineate critical signal transduction events involved in differentiation of hematopoietic, melanoma, and breast cell lineages and has over 50 scientific publications and 4 patents.

Collart manages a research program for DOE that focuses on the development of genome-scale methods for cloning and expression of proteins from the genomes of *Shewanella oneidensis* and *Geobacter sulfurreducens*. These organisms have potential for degrading organic pollutants and bioremediating metals. The program uses in vivo and cell-free approaches to address protein classes that represent a challenge for current cellular expression systems but are essential experimental targets for DOE research programs. With colleagues at MCSG, he has developed automated protocols for high-throughput generation and analysis of bacterial expression clones.

STEVE COLSON

Steve Colson received his bachelor's in chemistry from Utah State University in 1963 and his doctorate from the California Institute of Technology in 1968. He became associate laboratory director of the Fundamental Science Directorate in 2003 when he joined the leadership team at Pacific Northwest National Laboratory. Directorate divisions include Atmospheric Sciences and Global Change, Chemical Sciences, and Biological Sciences.

Colson has published more than 130 papers in peer-reviewed journals and has one patent. Before moving to PNNL, he spent 21 years at Yale University as a professor of chemistry. He research focused on the combination of optical and mass spectrometric methods to address fundamental problems in physical and analytical chemistry. General research interests include high-sensitivity spectroscopy and microscopy, photochemistry, photophysics, molecular dynamics, electronic structures of molecules, radical and molecular ions, process at the molecule-surface interface, and intermolecular interactions in molecular solids.

Before leaving Yale, he built up a strong collaborative team of faculty from the chemistry, physics, and engineering departments and industry. The power and excitement of interdisciplinary, collaborative research led him naturally to join the team created to establish the W. R. Wiley Environmental Molecular Sciences Laboratory, with a focus on the integration of modern physical, biological, and computational sciences.

MICHAEL COTTA

Michael Cotta is research leader for the Fermentation Biotechnology Research Unit (FBT), U.S. Department of Agriculture (USDA) Agricultural Research Service (ARS), National Center for Agricultural Utilization Research, in Peoria, Illinois. FBT conducts a broad-based program of microbial, biochemical, genetic, and engineering research to develop bioproducts and bioprocesses for conversion of agricultural commodities into biofuels and chemicals, enzymes, and polymers.

Cotta obtained his bachelor's and master's degrees in animal science in 1977 and 1979, respectively, from the University of California, Davis, where he worked under the direction of R. L. Baldwin. Upon completion of these studies, he continued his education at the University of Illinois in the laboratories of R. B. Hespell and M. P. Bryant. Cotta earned a doctorate in dairy science in 1985 and joined USDA ARS as a research microbiologist in October 1984. His research interests include microbial ecology of gastrointestinal environments and animal waste–handling systems, ecophysiology of ruminal microorganisms, and microorganism interactions in the bioconversion of complex polysaccharides.

BRUCE E. DALE

Bruce Dale is professor of chemical engineering and former chair of the Department of Chemical Engineering and Materials Science at Michigan State University (MSU). He earned his bachelor's degree (summa cum laude) in chemical engineering from the University of Arizona, Tucson, in 1974 and master's degree from the same university in 1976. He then studied under George T. Tsao at Purdue University, receiving his doctorate in 1979. His first academic position was in the Department of Agricultural and Chemical Engineering at Colorado State University, where he rose to professor in 1988. In that same year, he joined Texas A&M University, where he was professor of chemical engineering and of agricultural engineering. He also directed two multimillion-dollar interdisciplinary research centers at Texas A&M: Engineering Biosciences Research Center and the Food Protein Research and Development Center.

In 1996 Dale became professor and chair of the Department of Chemical Engineering at MSU, where he also holds an appointment in the Michigan Agricultural Experiment Station. In 1996 he won the Charles D. Scott Award for contributions to the use of biotechnology to produce fuels and chemical and other industrial products from renewable plant resources. In 2001 he stepped down as department chair to return to full-time research and teaching.

His research and professional interests lie at the intersection of chemical engineering and the life sciences. Specifically, he is interested in the environmentally sustainable conversion of plant matter to industrial products—fuels, chemicals, and materials—while meeting human and animal needs for food and feed. Dale expects to devote the rest of his MSU career to teaching and research aimed at developing such resources while the Hydrocarbon Age is winding down during the current century. His concern with sustainable resources was influenced by growing up in the copper-mining town of Ruth in eastern Nevada—a vibrant small community that became a ghost town when the mine ran out. Dale has distrusted societies that rely on mining natural resources (petroleum, for example) ever since.

He led production of the May 2000 National Research Council report, *Biobased Industrial Products: Research and Commercialization Priorities*. He has authored more than 100 refereed journal papers and is an active consultant to industry and an expert witness. He also holds 13 U.S. and foreign patents.

BRIAN H. DAVISON

Brian H. Davison is chief scientist for systems biology and biotechnology at Oak Ridge National Laboratory (ORNL), where for 2 years he was director of the Life Sciences Division. He previously was a distinguished researcher and leader of the Biochemical Engineering Research Group. In his 20 years at ORNL, he has conducted biotechnology research in a variety of areas, including bioconversion of renewable resources (ethanol, organic acids, and solvents); nonaqueous biocatalysis; systems analysis of microbes (cultivation and proteomics); biofiltration of volatile organic compounds; mixed cultures; immobilization of microbes and enzymes; metal biosorption; and extractive fermentations. His research has resulted in more than 80 publications and 6 patents.

Davison received his doctorate in chemical engineering from the California Institute of Technology and his bachelor's in chemical engineering from the University of Rochester. He chaired the 15th to 26th Symposia on Biotechnology for Fuels and Chemicals and served as proceedings editor in *Applied Biochemistry and Biotechnology* from 1994 to 2005. The symposium grew from 150 to 400 attendees during his 12 years as chair. Davison received an R&D 100 Award in 1977 for "Production of Chemicals from Biologically Derived Succinic

Acid." He also is an adjunct professor of chemical engineering at the University of Tennessee, Knoxville.

BILL DEAN

Bill Dean received his doctorate in biochemistry from Syracuse University. He is vice president of development and process sciences at Danisco Genencor International, Inc.; previously, he held the positions of vice president of technology programs and vice president of manufacturing development. Before joining Genencor, he was in the Research Division of Corning Glass Works, where he worked on enzyme immobilization techniques and bioreactor design and was responsible for overseeing the enzyme subcontract with the National Renewable Energy Laboratory to develop cost-effective biomass cellulases. Currently, he is responsible for grain-processing technology at Genencor.

TIM DONOHUE

Tim Donohue is professor of bacteriology at the University of Wisconsin, Madison. He earned his doctorate from Pennsylvania State University in 1980. He studies photosynthetic bacteria that convert solar energy into alternative fuels (hydrogen) or remove greenhouse gases and other environmental pollutants. He has used molecular genetic, biochemical, and systems biology techniques to study global signal-transduction pathways, alternative sigma factors, and signals that control expression of well-studied components of the respiratory and photosynthetic electron-transport chains.

Donohue's most recent work includes identification of cellular pathways used by photosynthetic microbes to sense the presence of singlet oxygen and defend themselves from this toxic substance. This knowledge may lead to the ability to fine tune the design of microbial and plant photosynthetic systems to minimize the harmful effects of singlet oxygen and to enhance energy production. Photosynthesis provides >90% of net energy input into the biosphere. Therefore, light-driven processes within photosynthetic organisms have enormous capacity for the production of sustainable, carbon-neutral, solar-powered technologies that reduce the global dependency on fossil fuels.

The long-range goals of Donohue's projects are to identify important metabolic and regulatory activities; obtain a thorough understanding of energy-generating pathways of agricultural, environmental and medical importance; and generate computational models to help design microbial machines with increased capacity to use solar energy, generate renewable sources of energy, remove toxic compounds, or synthesize biodegradable polymers.

TOM FOUST

Tom Foust joined the National Renewable Energy Laboratory (NREL) in 2004 as director of biomass research. He has a doctorate from the University of Idaho, master's from

Johns Hopkins University, and bachelor's from Pennsylvania State University, all in mechanical engineering. He also is a licensed professional engineer.

In his current role, he guides and directs NREL's research efforts to develop biomass conversion technology via both bioconversion and thermoconversion. This research is focused on developing the necessary science and technology for converting biomass to biofuels in an economical manner and covers the gamut of fundamental to applied science. His particular area of expertise is in complex flow and chemical-reaction modeling as it relates to biomass-conversion processes and in-process separations.

Before joining NREL, Foust spent 7 years with the Idaho National Laboratory where he was the research lead for the biomass feedstocks program. His primary area of research was in complex multiphase flow analysis as it related to physical fractionation of biomass. He has over 20 years of experience in research and research management, specializing in biomass feedstocks and conversion research. He has written more than 15 peer-reviewed publications related to biomass fractionation and technology-development issues.

JIM FREDRICKSON

Jim Fredrickson obtained a bachelor's in soil science from the University of Wisconsin, Stevens Point, and advanced degrees in soil chemistry and soil microbiology from Washington State University. He is a chief scientist within the Biological Sciences Division at Pacific Northwest National Laboratory, specializing in microbial ecology and environmental microbiology. With his research focused on subsurface microbiology and biogeochemistry, he has been responsible for laboratory and field research programs investigating the microbial ecology and biogeochemistry of geologically diverse subsurface environments and is recognized nationally and internationally for these studies.

Fredrickson also has served as subprogram coordinator for DOE's Subsurface Science Program from 1991 to present. In this role, he coordinated the technical aspects of DOE's Deep Subsurface Microbiology Subprogram at the national level and assisted the program manager in setting programmatic research directions. This subprogram involved more than 15 projects at universities and national laboratories nationwide and focused on multidisciplinary field-scale research. At the request of DOE, he currently is national coordinator for the *Shewanella* Microbial Cell Project, part of the Genomics:GTL program. He was appointed chief scientist in 2005 to serve as spokesperson to the science community for the GTL program and facilities.

RAMON GONZALEZ

Ramon Gonzalez is the William W. Akers Professor in the Department of Chemical and Biomolecular Engineering at Rice University. He holds degrees in chemical engineering from the University of Chile and the Central University of Las Villas, Cuba, and also is a licensed professional engineer. His research addresses such challenging issues in microbial catalysis as understanding and manipulating vitamins and cofactor biosynthesis, anaerobic fermentation of nontraditional carbon sources, simultaneous metabolism of sugars in sugar mixtures, and understanding and modifying respiratory and fermentative systems for synthesis of reduced products. Specific research areas include metabolic engineering and inverse (metabolic) engineering, functional genomics and systems biology, microbial fermentation, molecular modeling, and high-performance liquid chromatography optimization.

Gonzalez uses a wide spectrum of approaches and state-of-the-art techniques typically viewed under such different scientific and engineering disciplines as molecular biology, biochemistry, and chemical engineering. He currently is using transcriptomics and proteomic tools in conjunction with fluxomic tools to elucidate biological function of individual genes at cellular levels. He advocates the systemic method for its integration of mathematical and computational tools and is using this global and integrative approach to understand complex metabolic and regulatory networks in bacterial systems, the basis for understanding similar processes in more complex organisms. The ultimate goal of his research is the design of specific genotypes based on the desired phenotype.

BONNIE HAMES

Bonnie Hames leads the biomass chemical characterization teams within the National Bioenergy Center at the National Renewable Energy Laboratory (NREL). She earned a bachelor's degree in chemistry from Regis University and a doctorate in organic chemistry from the University of Denver. Working in the group of Bernard Monties, she also completed a postdoctoral assignment in lignin chemistry at the Centre National de la Recherche Scientifique, Institute National de la Recherche Agronomique, Laboratoire de Chimie Biologique, ThivervalGrignon-Paris, France.

Her extensive experience in biomass chemistry includes more than 18 years of developing standard wet chemical methods for characterizing biomass feedstock and biomass-derived materials, preparing standard reference materials for quality assurance and quality control, and applying standard methods to fuel and chemical production from biomass. She currently leads the NREL Biomass Program to develop new, rapid, and inexpensive methods for biomass compositional analysis. These methods include advanced tools for chemical characterization of biomass feedstocks and biomass-derived materials based on infrared spectroscopy and advanced multivariate analysis techniques first developed in the Agenda 2020 program sponsored jointly by DOE Office of Industrial Technologies and the forest products industry. In 2000, her program's real-time biomass analysis won an R&D 100 award and was honored by DOE as a Best of Agenda 2020 Project.

Hames also developed patented techniques for biomass fractionation and lignin isolation. She has extensive experience in lignin chemistry including structural characterization, synthesis of lignin model compounds, and development of methods for the selective oxidation of lignin using organ metallic catalysts and biomimetic systems. She holds 3 U.S. patents, has authored 3 book chapters and more than 25 papers in peer-reviewed journals, and has made more than 80 presentations at technical meetings. She currently chairs the American Society for Testing and Materials committee E48 on standards for biotechnology and subcommittee E48.05 on standards of biomass conversion.

MARIA HARRISON

Maria Harrison earned her bachelor's degree with honors in microbiology from the University of Newcastle Upon Tyne, England, and her doctorate in 1987 from the Institute of Science and Technology, University of Manchester. She conducted postdoctoral research at the Samuel Roberts Noble Foundation in Ardmore, Oklahoma, under the direction of R. A. Dixon. She has served as an adjunct professor at Oklahoma State University and Texas A&M University; in 2003, she joined the staff at the Boyce Thompson Institute for Plant Research, with an adjunct appointment in the Department of Plant Pathology at Cornell University.

Most vascular flowering plants are able to form symbiotic associations with arbuscular mycorrhizal (AM) fungi. These associations develop in the roots, where the fungus colonizes cortical cells to access carbon supplied by the plant. The fungal contribution to symbiosis includes transfer of mineral nutrients, particularly phosphorus, from the soil to the plant. In many soils, phosphate levels are limiting to plant growth. Consequently, additional phosphate supplied via AM fungi can have a significant impact on plant development, and this symbiosis influences the structure of plant communities in ecosystems worldwide.

The long-term goals of Harrison's research are to understand the mechanisms underlying development of AM symbiosis and phosphate transfer among symbionts. She uses a model legume *Medicago truncatula* and AM fungi *Glomus versiforme*, *G. intraradices,* and *Gigaspora gigantea* for these analyses. Currently, a combination of molecular, cell biology, genetic, and genomic approaches is being used to obtain insights into symbiosis development, communication among plant and fungal symbionts, and symbiotic phosphate transport.

SUSAN HENNESSEY

In almost 20 years at DuPont, Susan Hennessey has applied her expertise as a chemical engineer to a number of improved chemical production processes. Her team of scientists has used immobilized bacterial cells and other encapsulation techniques to increase enzyme durability and concentration for continuous production. Her work on the introduction of the herbicide Milestone for crop protection earned her team the 2002 Industrial Innovation Award from the Mid-Atlantic Region of the American Chemical Society. Other work combining the synergy of biology with new chemical synthesis technologies continues to result in new high-value products using environmentally preferred processes.

MIKE HIMMEL

During his 23-year term at DOE's National Renewable Energy Laboratory (NREL, formerly SERI) in Golden, Colorado, Mike Himmel has worked to support many technical aspects of DOE's Biomass Program. More recently, he has been responsible for establishing the facilities and staff necessary to pursue projects in protein engineering, specifically cellulases. Today, he manages the award-winning Enzyme Technology Team and the major industrial and academic subcontracts that support this work. The team has assembled world-class protein purification and characterization facilities at NREL, with special emphasis on robotic systems for screening libraries derived from directed evolution technology. Himmel has contributed 300 peer-reviewed papers and meeting abstracts, 4 books, and 16 patents to the literature. He also chaired or cochaired 15 international meetings in the field of biochemistry and biotechnology, including the 2003 Gordon Research Conference on Cellulases and Cellulosomes.

MAURICE HLADIK

Maurice Hladik is director of marketing for Iogen Corporation, which specializes in developing, manufacturing, and marketing enzymes to modify and improve the processing of natural fibers within the textile, animal-feed, and pulp and paper industries. One of Iogen's major activities is research on cellulosic ethanol. Hladik has extensive international business experience, particularly in the United States, Germany, and the United Kingdom as well as several Asian countries including China (also Hong Kong), South Korea, and Thailand. A particular strength is his ability to locate business contacts and commercial intelligence in a foreign setting.

Before joining Iogen, from 1978 to 1998 he served in the Canadian Foreign Service. His assignments included senior Canadian trade officer posted to Bangkok (3 years), Hong Kong (3 years), Beijing (1 year), Seoul (2 years), and Munich as consul general (3 years); and director general of the Grain Marketing Bureau at Agriculture Canada. In this position, Hladik was chief advisor to the Canadian government on international grain policy and marketing issues. He also served the Canadian International Development Agency as director for Asia, Industrial Cooperation Division.

LONNIE INGRAM

Lonnie Ingram is a distinguished professor of microbiology and director of the Florida Center for Renewable Chemicals and Fuels at the University of Florida. Elected to the National Academy of Sciences in 2001, Ingram and his coworkers were the first in the world to develop genetically engineered *E. coli* bacteria capable of converting all sugar types found in plant cell walls into fuel ethanol. Ingram's organism produces a high yield of ethanol from such biomass as sugarcane residues, rice hulls, forestry and wood wastes, and other organic materials.

Ingram's breakthrough bioconversion technology was selected to become Landmark Patent No. 5,000,000 by the U.S. Department of Commerce. More than 30 additional patents are pending or have been issued for this technology, which is being commercialized with assistance from the U.S. Department of Energy. BC International Corp., based in Dedham, Mass., holds exclusive rights to use and license the engineered bacteria, dubbed "KO11" by Ingram. In 1993, Ingram received a U.S. Department of Agriculture Distinguished Service Award for his breakthrough research. The agency's highest honor, the award recognized his outstanding contributions to research and the consumer.

JAY KEASLING

Jay Keasling earned a bachelor's degree in chemistry and biology in 1986 from the University of Nebraska, Lincoln, and a doctorate in chemical engineering from the University of Michigan in1991. He is professor of chemical engineering and bioengineering at the University of California, Berkeley. He also serves as director of the Physical Biosciences Division and heads the new Synthetic Biology Department at Lawrence Berkeley National Laboratory. The idea behind this department is to design and construct novel organisms and biologically inspired systems to solve problems unsolvable by natural biological systems and also provide new information about living cells. Keasling's many honors include election as fellow of the American Institute of Medical and Biological Engineering and recipient of the AIChE Award for Chemical Engineering Excellence in Academic Teaching. He also is founder of two companies, Amyris Biotechnologies and Codon Devices, which have grown from discoveries at his laboratory.

Keasling and collaborators were awarded a Gates Foundation grant that will seek to create in the laboratory an inexpensive antimalarial drug, artemisinin. This drug could be sold for one-tenth of today's price and compete with the formerly front-line antimalarial now confronted by disease-resistant strains around the world. Keasling and his team at Berkeley already have worked out methods for extracting the genes responsible for making artemisinin and have transplanted them into a harmless strain of *E. coli*.

Research in the Keasling laboratory focuses on the metabolic engineering of microorganisms for degradation of environmental contaminants or for environmentally friendly synthesis. To that end, he has developed a number of new genetic and mathematical tools to allow more precise and reproducible control of metabolism. These tools are being used in such applications as synthesis of terpene drugs and biodegradable polymers, accumulation of phosphate and heavy metals, degradation of chlorinated and aromatic hydrocarbons, biodesulfurization of fossil fuels, and complete mineralization of organophosphate nerve agents and pesticides. Genomics, proteomics, and metabolomics are being employed to investigate effects of these changes on cellular physiology and to optimize cellular redesign.

KEN KEEGSTRA

Ken Keegstra is director of the Plant Research Laboratory and a University Distinguished Professor in the departments of Biochemistry and Plant Biology at Michigan State University (MSU). He received his doctorate in chemistry at the University of Colorado, where he investigated the structure of plant cell-wall components and their interactions within the wall. For more than 20 years he studied other biological problems, mainly chloroplast biogenesis and the targeting of nuclear-encoded protein into chloroplasts. At the time of his move to MSU in 1993, he reinitiated work on plant cell walls. The major focus of his current research is the biosynthesis of plant cell-wall polysaccharides produced in the Golgi before delivery to the cell wall. The greatest amount of effort has centered on xyloglucan biosynthesis, but his research group also has investigated the biosynthesis of mannans, glucomannans, and arabinoxylans as well as a few other wall components.

MICHAEL LADISCH

Michael Ladisch is director of the Laboratory of Renewable Resources Engineering and Distinguished Professor of Agricultural and Biological Engineering at Purdue University, with a joint appointment in biomedical engineering and a courtesy appointment in food science. In 1973 he earned his bachelor's degree from Drexel University and in 1974 and 1977, respectively, his master's and doctorate from Purdue University, all in chemical engineering.

He has a broad background in bioscience and bioengineering and has authored a textbook, *Bioseparations Engineering: Principles, Practice and Economics* (Wiley, 2001) and 150 journal and proceedings papers. He has 14 patents (issued and applied for) and has presented over 100 papers. He received the Marvin J. Johnson Award in Biochemical Technology of the American Chemical Society in 2002 and the Food, Pharmaceutical, and Bioengineering Division Award of the American Institute of Chemical Engineers in 2001. He was elected to the National Academy of Engineering in 1999.

Ladisch served as a member of U.S. delegations and advisory panels to Russia, Thailand, China, and Japan to review the status of biotechnology programs. In 1991 and 1992, he chaired the National Research Council's Committee on Bioprocess Engineering, which studied research priorities and policy issues related to commercialization of biotechnology and published the report *Putting Biotechnology to Work: Bioprocess Engineering* (National Academy Press, 1992).

His research addresses fundamental topics in bioprocess engineering as they apply to bioenergy, bioproducts, biorecovery, and bionanotechnology. The research addresses transformation of renewable resources into bioproducts, properties of proteins and living organisms at surfaces, rapid prototyping of microfluidic biosensors, and bioseparations. This work has resulted in new industrial bioenergy processes and systematic approaches and correlations for scaleup of laboratory chromatographic purification techniques to process-scale manufacturing systems; it also has resulted in scaledown of bioseparations and the rapid prototyping of microfluidic biochips for quick detection of pathogenic microorganisms. Ladisch teaches bioseparations, bioprocess engineering, and biotechnology at both the graduate and undergraduate levels.

LEE LYND

Lee Rybeck Lynd is a professor of engineering and adjunct professor of biological sciences at Dartmouth College and a professor extraordinary of microbiology at the University of Stellenbosch in South Africa. He received a bachelor's degree in biology from Bates College; master's in bacteriology from the University of Wisconsin, Madison; and master's and doctorate from the Thayer School of Engineering.

Lynd leads a research group in biochemical engineering and applied biology relevant to processing cellulosic biomass. His laboratory's research topics are chosen to address a primary technical impediment to realizing a "carbohydrate economy": Overcoming the recalcitrance of cellulosic materials to biological conversion. Specific topics include, in order of increasing scale, metabolic engineering to improve product yields in thermophilic bacteria, microbial physiology of anaerobic cellulolytic microorganisms, kinetics and reactor design for enzymatic and microbial hydrolysis of cellulosic materials, and conversion of "real-world" cellulosic materials such as waste sludge produced from paper mills. A particular focus of the Lynd group is "consolidated bioprocessing," a widely applicable potential breakthrough in which production of cellulase enzymes, hydrolysis of biomass fiber, and fermentation of resulting sugars are accomplished in one process step by a single microbial community.

Lynd is a recipient of the National Science Foundation's Presidential Young Investigator Award and a two-time recipient of the Charles A. Lindbergh Award for his efforts to promote balance between technological progress and preservation of natural and human environments. Professional activities include service as associate editor for *Biotechnology and Bioengineering*, member of a presidential advisory committee on reducing greenhouse gas emissions from personal vehicles, and organizing committee member for the Annual Symposium on Biotechnology for Fuels and Chemicals. Lynd has authored more than 60 peer-reviewed manuscripts and holds 5 patents.

GEORGE MICHAELS

In May 2004, George Michaels was named associate laboratory director of the newly formed Computational and Information Sciences Directorate at Pacific Northwest National Laboratory (PNNL). This new directorate delivers innovative solutions to address national and global problems by enabling large-scale scientific discoveries through R&D in science-driven computing. He holds a doctorate in biochemistry and molecular biology and a bachelor's degree (1974) in microbiology, all from the University of Florida.

He joined PNNL in April 2003 as director of bioinformatics for the Biomolecular Systems Initiative. He is an internationally recognized pioneer in bioinformatics and in the practical development of biotechnological approaches for discovery. During his career spanning nearly 30 years, he has provided increasingly significant technical and leadership contributions to his field. He holds patents in methods for designing DNA-binding proteins and for morphological reconstruction.

Most recently, Michaels held leadership positions at Monsanto in St. Louis, Missouri, where he designed an integrated expression-profiling program. He also cofounded and served as vice- president and chief scientist of Genome Dynamics, a Maryland biotechnology startup company. While an associate professor at George Mason University in Fairfax, Virginia, he initiated one of the nation's first doctoral programs in bioinformatics and computational biology. He also has served as a special expert to the office of the director of the National Institutes of Health.

COLIN MITCHINSON

Colin Mitchinson earned his bachelor's degree in biochemistry from the University of Edinburgh and doctorate in biochemistry from the University of Newcastle. He has extensive research experience in the study of structure-function relationships in proteins ranging from

(Ca++/Mg+) ATPase of muscle sarcoplasmic reticulum, ribonuclease, and subtilisin. He also has performed protein engineering on starch-processing enzymes and cellulases and was project leader and principal investigator for a multidisciplinary effort to develop a new cellulase. He currently serves as senior staff scientist at Danisco Genencor, where his research focuses on development of new cellulase products for biomass conversion. His publications include reviews on protein folding, substrate binding, and active site characterization using molecular genetic, biochemical, and biophysical techniques. A representative recent publication is M. Sandgren, J. Stahlberg, and C. Mitchinson, "Structural and Biochemical Studies of GH Family 12 Cellulases: Improved Thermal Stability, and Ligand Complexes," *Progress in Biophysics and Molecular Biology* **89**, 246–91 (2005).

MARK MORRISON

Mark Morrison is a professor in the Department of Animal Science at Ohio State University and also holds a nonsalaried appointment in the Department of Microbiology. He has a long-standing interest in gastrointestinal microbiology and bacterial physiology. Much of his research has focused on the ecophysiology of plant biomass degradation in herbivores and the molecular biology underpinning cellulose degradation and bacterial adhesion to plant structural polysaccharides. He serves as project leader for the North American Consortium for Genomics of Fibrolytic Bacteria, involving scientists from The Institute for Genomic Research, Cornell, University of Illinois, and University of Guelph. The consortium's activities include sequencing of four rumen bacterial genomes (*Fibrobacter succinogenes*, *Prevotella bryantii*, *Prevotella ruminicola*, and *Ruminococcus albus*) as well as comparative and functional genomic studies with these bacteria.

ART RAGAUSKAS

Art Ragauskas received his doctorate in chemistry from the University of Western Ontario in 1986, with subsequent postdoctoral research at the University of Alberta and Colorado State University. He is a fellow of the International Academy of Wood Science and the Technical Association of the Pulp and Paper Industry (known as TAPPI). His research program at Georgia Institute of Technology is seeking to understand and exploit innovative sustainable lignocellulosic materials and develop new and improved applications for nature's premiere renewable biopolymers including cellulose, hemicellulose, and lignin.

Ragauskas's research is directed toward innovative processes for converting lignocellulosic biomass into innovative biomaterials and biofuels. Achieving this goal requires research in several fields of study, including lignocellulosic fiber chemistry and physical properties; carbohydrate, lignin, and extractive chemistry; nanobiomaterials; biotechnology; and material and polymer science. These studies are supported by expertise in advanced spectroscopy, imaging, nanotechnology, chemoenzymatic biotechnology, cold plasma, composites, bleaching, and pulping technologies. His research is sponsored by a consortium of industrial partners, Defense Advanced Research Projects Agency, National

Science Foundation, U.S. Department of Agriculture, Department of Energy, and Georgia Traditional Industries Program (TIP3).

Ragauskus has been a Luso-American Foundation Teaching Fellow at the Universidade da Beira Interior, Portugal; an invited guest teaching professor at Chalmers University of Technology, Sweden and South China University of Technology; and an invited research professor at Royal Institute of Technology, Stockholm. He has published 185 papers, patents, and conference proceedings. He is an associate editor for the *Journal of Pulp and Paper Science*, Holzforschung, and the *Journal of Chemistry and Technology* and has served on several advisory boards and review panels.

JOHN RALPH

John Ralph received his bachelor's degree with honors in chemistry in 1976 from the University of Canterbury, New Zealand, and his doctorate in chemistry and forestry in 1985 from the University of Wisconsin, Madison (UWM). He is a research chemist at the U.S. Dairy Forage Research Center and a professor in the UWM Department of Forest Ecology and Management.

An organic chemist and nuclear magnetic resonance (NMR) spectroscopist specializing in cell-wall model compound syntheses and monolignol polymerization reactions, Ralph is involved principally in studies aimed at detailing the mechanisms of lignin-polysaccharide cross-linking and their effect on limiting cell-wall degradability. He developed NMR methods for cell-wall structural analysis, including an NMR database of model compounds for lignin and related wall components, as well as methods for analyzing lignin structure and nondegradatively solubilizing the entire cell-wall faction of finely divided plant cell walls. He has additional expertise in synthetic organic chemistry, specifically in the synthesis of cell-wall model compounds, lignin oligomers, enzyme precursors, and products.

SIMO SARKANEN

Simo Sarkanen is a professor in lignin chemistry and biochemistry at the Department of Biobased Products, University of Minnesota. He received his undergraduate training at King's College, Cambridge (England), and was awarded a doctorate in chemistry from the University of Washington, Seattle. His first publications were in theoretical (computational) chemistry, but his doctoral dissertation was in bioorganic chemistry (enzyme kinetics). At the postdoctoral level in the Department of Chemical Engineering at the University of Washington, he embarked on a journey into various controversial aspects of lignin chemistry. Currently, his research interests range from lignin biosynthesis and biodegradation to new formulations for lignin-based thermoplastics.

Most lignin chemists and biochemists have thought that configurations of lignin macromolecules are random (or combinatorial), but Sarkanen's group is trying to develop an explicit working hypothesis for replicating specific lignin primary structures during lignin biosynthesis. The first step in lignin biodegradation generally has been considered under the control of lignin peroxidase, manganese-dependent peroxidase, or laccasemediator systems.

Sakanen and his coworkers actively promote the view that a completely different kind of lignin depolymerase may be responsible for cleaving lignin macromolecules in vivo. Finally, they have produced the first series of thermoplastics with promising mechanical properties composed predominantly or entirely (85 to 100%) of simple lignin derivatives. Previous work in the field typically had encountered incorporation limits of 25 to 40% for lignins in potentially useful polymeric materials.

CHRISTOPHE SCHILLING

Christophe Schilling is a cofounder of Genomatica, Inc. He received his doctoral degree in bioengineering under Bernhard Palsson at the University of California, San Diego, where he was a Powell Foundation and Whitaker Foundation Fellow. He also holds a bachelor's degree in bomedical engineering from Duke University, where he was a Howard Hughes Undergraduate Research Fellow.

As president and chief scientific officer of Genomatica, he currently directs all the company's business and scientific efforts toward applying advanced modeling and simulation technologies to a number of metabolism-driven products. This includes overseeing development of Genomatica's integrated computation and experimental platform to drive the metabolic engineering of microbes to support next-generation bioprocesses being developed by the company's industrial partners.

He is coauthor of numerous scientific articles in systems biology and metabolic modeling, and he is an inventor on a number of patent filings surrounding Genomatica's core technologies. He was featured in the February 2001 edition of *Genome Technology* as one of 16 "up and comers" who have the talent and drive to make great strides in science, technology, and business. In 2003 he was named one of the top 100 young innovators under 35 whose innovative work in technology will have a profound impact on the world, as selected by MIT's *Technology Review* magazine.

JOHN SHANKLIN

John Shanklin is a senior biochemist in the Biology Department at Brookhaven National Laboratory and an adjunct professor in the Biochemistry Department at the State University of New York at Stony Brook. He received his bachelor's degree in physiology from the University of Lancaster, United Kingdom, in 1981 and his master's in forestry from the University of Wisconsin, Madison (UWM), in 1984. He was awarded a doctorate in horticulture from UMW in 1988, working on the ubiquitin system.

Shanklin's current interests focus on plant lipid biochemistry and developing plant oils as renewable industrial feedstocks. Specifically he studies structure and function in lipid modification enzymes. This class of enzymes performs high-energy chemistry on a wide variety of substrates with various chemical outcomes. In addition to studying how specific fatty acids are synthesized, he also conducts metabolic engineering experiments on how modulations in pathway components affect storage oil accumulation for both usual and unusual fatty acids.

Shanklin is the winner of several awards, including the Office of Energy Research Young Scientist Award, a Presidential Early Career Award, and the Terry Galliard Medal for Plant Lipid Biochemistry. He was chair of the Scientific Program Committee for the National Plant Lipid Cooperative meetings from 2001 to present. He has served on panels for the Department of Energy and the National Science Foundation and is on the Scientific Advisory Board for *Genetic Engineering, Principles and Methods* and *Advances in Plant Biochemistry and Molecular Biology*. He currently is a member of the Council for Energy Biosciences on the Basic Energy Sciences Committee of Visitors (2005).

SHARON SHOEMAKER

Sharon Shoemaker joined the University of California, Davis (UCD), in 1991 as founder and executive director of the California Institute of Food and Agricultural Research (known as CIFAR). She also is research leader at the UC Forest Products Laboratory and site director of the National Science Foundation–UCD Center for Advanced Processing and Packaging. Shoemaker holds a bachelor's degree in chemistry, master's in food science, and doctorate in biochemistry and nutrition from Virginia Tech. After postdoctoral training in biochemical engineering, she joined Cetus Corporation in Berkeley, California, the first U.S. biotechnology company. After 7 years, she left Cetus to join Genencor, a company more aligned with her interests in applying fermentation and enzymology to problems in food and agriculture.

Shoemaker's industrial experience led to patents on novel yeast strains to convert biomass to ethanol and on novel bacterial strains to produce new forms of cellulose. She also led team efforts to improve enzyme systems for converting biomass to sugars and subsequently fermenting them to chemicals and to characterize and develop ligninase systems for use in the pulp and paper industry. Her research interests focus on cellulose applications in biomass conversion (e.g., rice straw, wood, and mixed waste paper); integration of various unit operations in biomass-conversion processes (membrane filtration and enzymes); and development of new analytical methods for quantifying specific cellulase activities. Shoemaker is active in regional, national, and international task forces, review panels, and programming on new and emerging biobased processing technologies, carbon sequestration, and cellulase R&D.

YUVAL SHOHAM

Yuval Shoham has been head since 2004 of the Department of Biotechnology and Food Engineering at the Israel Institute of Technology (Technion) in Haifa. He received his bachelor's degree in biology from Tel Aviv University in 1980, his master's in microbiology in 1982, and his doctorate in biochemical engineering from MIT in 1987. In 1988 he joined the Technion, where he is director of the Otto Meyerhoff Minerva Center for Biotechnology and holds the Erwin and Rosl Pollak Chair in Biotechnology. He is a fellow of the American Academy of Microbiology.

Shoham's research focuses on the catalytic mechanisms and structure-function relationships of industrial enzymes, especially glycoside hydrolases, and on gene regulation of the hemicellulolytic system in *Geobacillus stearothermophilus* and cellulosomerelated genes in *Clostridium thermocellum*. He has been involved in several industrial projects, including the development of a large- scale process for bleaching paper pulp with alkaline thermostable xylanases and an enzymatic process for making complex lipids with novel SN-2 lipases. He has authored more than 120 articles and book chapters and holds several patents.

LLOYD M. SMITH

Lloyd M. Smith is John D. MacArthur Professor of Chemistry and director of the Genome Center at the University of Wisconsin, Madison (UWM), where he has been since 1988. He received a bachelor's degree in biochemistry from the University of California, Berkeley, in 1977 and a doctorate in biophysics from Stanford University in 1981. In 1982 he moved to the California Institute of Technology, where he developed the first fluorescence-based automated DNA sequencing instrument.

Smith has been named one of *Science Digest*'s Top 100 Innovators and has received the Presidential Young Investigator Award, Eli Lilly Analytical Chemistry Award, Association of Biomolecular Resource Facilities Award for the development of automated DNA sequencing, and the American Chemical Society Award in Chemical Instrumentation. He has served on the NIH National Human Genome Research Institute Advisory Council and the NIH Human Genome Study Section, has authored more than 165 scientific papers, and is inventor on 20 issued U.S. patents. He is a cofounder of the biotechnology company, Third Wave Technologies, and is a member of the board of directors of GWC Technologies, Inc., and GenTel Biosurfaces,Inc., where he also is chair of the Scientific Advisory Board. His primary area of research is the development of new technologies for analysis and manipulation of biomolecules.

CHRIS SOMERVILLE

Chris Somerville is director of the Carnegie Institution Department of Plant Biology and professor in the Department of Biological Sciences at Stanford University. He has published more than 170 scientific papers and received patents in plant and microbial genetics, genomics, biochemistry, and biotechnology. His current research interests are focused on the characterization of proteins such as cellulose synthase, which is implicated in plant cell-wall synthesis and modification. He is a member of the senior editorial committee of *Science* magazine and of the scientific advisory boards of numerous academic institutions and private foundations in Europe and North America. He is a member of the U.S. National Academy of Sciences, Royal Society of London, and Royal Society of Canada. He has received numerous scientific awards and several honorary degrees. He is chairman of the board of Mendel Biotechnology, a private plant biotechnology company in the San Francisco Bay area.

GREGORY STEPHANOPOULOS

Since 1985, Gregory Stephanopoulos has been a professor of chemical engineering at Massachusetts Institute of Technology (MIT). He received his bachelor's degree from the National Technical University of Athens, master's from the University of Florida, and doctorate from the University of Minnesota, all in chemical engineering. Upon finishing his doctorate in 1978, he joined the faculty of the California Institute of Technology, where he served as assistant and associate professor until 1985. He was associate director of the Biotechnology Process Engineering Center between 1990 and 1997 and was appointed Bayer Professor of Chemical Engineering and Biotechnology. He also is the Taplin Professor of Health Sciences and Technology (2001–), instructor of bioengineering at Harvard Medical School (1997–), member of the international faculty of the Technical University of Denmark (2001–), and fellow of the Singapore–MIT Alliance (2000–).

Stephanopoulos's current research focuses on metabolic engineering and its applications to the production of biochemicals and specialty chemicals, the rigorous evaluation of cell physiology using advanced isotopic methods, the metabolism and physiology of mammalian cells with emphasis on obesity and diabetes, and bioinformatics and functional genomics whereby new genomicsbased technologies are applied to the elucidation of cell physiology and metabolic engineering. He has coauthored or coedited 5 books and published some 250 papers and 19 patents.

Stephanopoulos currently is editor-in-chief of the journal *Metabolic Engineering* and serves on the editorial boards of seven scientific journals. In 1992 he chaired the Food, Pharmaceutical, and Bioengineering Division of the American Institute of Chemical Engineers (AIChE) and was elected a founding fellow of the American Institute for Medical and Biological Engineering. In 2002 he received the Merck Award in Metabolic Engineering and was elected to the board of directors of AIChE. In 2003, he was elected to the National Academy of Engineering and in 2005 was awarded an honorary doctorate (doctor technices honoris causa) by the Technical University of Denmark.

He has taught a variety of undergraduate and graduate courses in the chemical engineering curricula at California Institute of Technology and MIT. He also has developed a number of new courses, including Metabolic Engineering, Metabolic and Cell Engineering, and, more recently, Bioinformatics. He coauthored the first textbook on metabolic engineering and has taught a number of biotechnology courses in the summer sessions since 1985. He introduced and directed two such courses, Metabolic Engineering (1995–99) and Bioinformatics (2000–).

BRUCE STONE

Bruce Stone received his bachelor's degree from the University of Melbourne in 1948 after majoring in chemistry and biochemistry. In 1951 he was seconded for training in mycology to the Commonwealth Mycological Institute, Kew, England. In 1952 he commenced doctoral studies in the Department of Biochemistry at University College, London. After graduating in 1954, he held postdoctoral appointments in Ottawa (National Research Council Fellow) and London (Imperial Chemical Industries Fellow). He returned to the Russell Grimwade School of Biochemistry, University of Melbourne, as a lecturer in 1958 and was appointed reader in agricultural biochemistry in 1966. From 1972 until his official retirement in 1995, he served as foundation professor of biochemistry at La Trobe University.

Stone twice held the position of dean of the School of Biological Sciences at La Trobe (1976–78, 1987-91), was president of the Australian Biochemical Society (1988–90), and was chairman of the Royal Australian Chemical Institute, Cereal Chemistry Division (1978–79). He currently is editor-in-chief of the *Journal of Cereal Science* and assistant director of the Australian Academy of Science and Technology's Crawford Fund, an organization supporting international agricultural research.

His major research interest in the chemistry and biochemistry of plant polysaccharides arose during his first appointment through an investigation of cellulose-breakdown enzymology. His specific studies on polysaccharide components of the cell walls of cereals and grasses encompassed their structure, biosynthesis, depolymerization, and their interactions with lignins and proteins. Their outcomes have been applied to the solution of agricultural and horticultural problems, especially in relation to cereal-grain quality and processing behavior and in human and ruminant nutrition. Stone has a special interest in the biology and chemistry of callose and related($1 \rightarrow 3$)-ß-D-glucans, and, with Adrienne Clarke, published a treatise on the subject in 1993. Currently he is investigating the biology and biosynthesis of a bacterial ($1 \rightarrow 3$)-ß-D-glucan, curdlan.

BOB TABITA

Bob Tabita is professor of microbiology and plant biology and an Ohio Eminent Scholar at Ohio State University. His doctoral work in the late Don Lundgren's laboratory at Syracuse University introduced him to the metabolism and biochemistry of autotrophic bacteria; his postdoctoral research was in Bruce McFadden's group in the Chemistry Department at Washington State University. A key experiment, in which he discovered that growth with a reduced electron donor upregulates Rubisco synthesis, has been the foundation for his entire career with photosynthetic bacteria, enabling studies on the enzymology of Rubisco and other pathway enzymes. His laboratory continues this strong interest in molecular regulation, biochemistry, and enzymology of carbon dioxide assimilation and the control of Rubisco synthesis.

All organisms require CO_2. It is used in many enzyme-catalyzed reactions in processes as important and varied as carbohydrate metabolism, lipid biosynthesis, and production of vital metabolic intermediates for the cell. With the realization that many microorganisms use CO_2 to elicit pathogenesis, CO_2 metabolism and its control are recognized as having great health relevance. Carbon dioxide also may be employed as the sole source of carbon by a large and diverse group of organisms. For this reason, CO_2 fixation is associated with global issues of agricultural productivity, carbon cycling, and industrial productivity. Carbon dioxide also is recognized as the chief greenhouse gas and has been implicated in general warming of the earth's biosphere. For all these reasons, research on various aspects of CO_2 fixation control, biochemistry, and ecology have attracted wide interest. Microbial systems studied in the Tabita laboratory include *R. palustris*, *C. tepidum*, and *R. sphaeroides*.

STEVE THOMAS

Steve Thomas's research expertise spans work on insect hormones at the University of California, Los Angeles, to plant biotechnology research at the ARCO Plant Cell Research Institute in Dublin, California. Through the National Renewable Energy Laboratory, Thomas was manager for the DOE Office of Fuel Development project entitled, "Production of Cellulases in Tobacco and Potato Plant Bioreactors." He has worked on the Sugar Processing Integration Task, surveying the compositional variability of geographically and genetically diverse corn stover residues to minimize risk associated with commercialization of biomass-conversion technology. He recently joined Ceres, Inc., as a principal scientist.

JERRY TUSKAN

Gerald A. Tuskan, a distinguished scientist in Oak Ridge National Laboratory (ORNL) Environmental Sciences Division, holds a bachelor's degree in forest management from Northern Arizona University; a master's in forest genetics from Mississippi State University; and a doctorate in genetics from Texas A&M University. He also is a research professor in the University of Tennessee's departments of Entomology, Plant Pathology, Plant Sciences, and Genome Sciences and Technology where he advises graduate students, interacts with departmental faculty, and provides guest lectures and graduate seminars. Before joining ORNL, he served as associate professor of horticulture and forestry at North Dakota State University and was an instructor in the Forest Science Department at Texas A&M University.

At ORNL, Tuskan is responsible for coordinating the DOE effort to sequence the *Populus* genome. This includes projects with carbon allocation and partitioning in woody plants as a means to enhance bioenergy conversion and carbon sequestration, genome-enabled discovery of carbon sequestration genes in poplar, environmental influences on wood chemistry and density of *Populus* and loblolly pine, formation of an international *Populus* genome consortium, and creation of a *Populus* postsequence science plan. His research helps identify genes associated with cell-wall chemistry, genetic mapping in *Populus*, particularly related to carbon allocation and partitioning, and the use of genomics information to accelerate domestication of *Populus*.

ED UBERBACHER

Ed Uberbacher is lead scientist for computational biology in the Life Sciences Division at Oak Ridge National Laboratory. He received a bachelor's degree from Johns Hopkins University in 1974 and a doctorate in chemistry from the University of Pennsylvania in 1979. He is the codeveloper of GRAIL, the first of many gene-finding programs created during the Human Genome Project. Next-generation algorithms such as Grail-Exp, which can use both EST and complete cDNA data, provide another level to the analysis of draft data.

As a member of the Computational Biology Institute, Uberbacher works within a collaborative environment that combines the expertise of biologists, computer scientists, and mathematicians with high-performance computing to create and provide tools and infrastructure to advance systems and computational biology. He is CEO of Genomix Corporation, a provider of analysis systems, information resources, and partnering opportunities to guide R&D teams through critical steps in pharmaceutical discovery and development. He also serves as a faculty member in computational biology and bioinformatics at the University of Tennessee–ORNL Graduate School of Genome Science and Technology.

FERNANDO VALLE

Fernando Valle is staff scientist for process science at Danisco Genencor International, with expertise in microbial pathway engineering for the biological production of enzymes, therapeutic proteins, peptides, antibiotics, vitamins, amino acids, and organic acids. His significant research accomplishments include the optimization and pathway engineering of an *E. coli* strain to transform glucose into 1, 3-propanediol, a monomer used in producing polyester fibers.

JOHN VOGEL

John Vogel's research program is focused on applying molecular biology to improve herbaceous energy crops. Major projects include using dsRNA-mediated gene silencing to reduce lignin content in switchgrass and studying the model grass *Brachypodium distachyon* to identify genes controlling cell-wall composition. Recent accomplishments include creation of transgenic switch- grass plants with silenced lignin biosynthetic genes, development of an *Agrobacterium*-mediated transformation method for *Brachypodium*, and sequencing of >20,000 *Brachypodium* ESTs.

He is a USDA Agricultural Research Service molecular biologist at the Western Regional Research Center in Albany, California. His experience includes extensive research in plant pathology as both an assistant professor at the University of California, Riverside, and postdoctoral fellow at the Carnegie Institution of Washington in Stanford, California. His graduate work focused on cytokinin and ethylene signaling in *Arabidopsis*.

KEN VOGEL

Kenneth Vogel is a research geneticist at the USDA Agricultural Research Service Wheat, Sorghum, and Forage Research Unit in Lincoln, Nebraska. He received his bachelor's and master's degrees from Colorado State University in 1965 and 1967 and a doctorate from the University of Nebraska in 1974.

His responsibilities are in research management and perennial grass breeding and genetics. His specific research interest is to develop improved perennial grasses, switchgrass in particular, for use on marginal lands of the Central Great Plains and mid-western states along with associated management practices. This research involves germplasm evaluation and characterization of native and introduced grasses and basic and applied genetic research in addition to applied breeding work. Since 1990, he has been conducting research to develop switchgrass into a biomass energy crop for these areas.

NICHOLAS WHEELER

Nicholas Wheeler received an undergraduate degree in forest sciences from the University of Washington in 1973, a master's in forest genetics from Michigan State University in 1974, and a doctorate in plant breeding and genetics from Wisconsin in 1981. He has worked as a tree breeder for the government of British Columbia for 4 yrs, as a scientist in the forest products industry for 22 yrs, and, more recently, as a private consultant in his own business (Molecular Tree Breeding Services).

Wheeler's research includes projects in physiological and quantitative genetics, genecology, and molecular genetics. He currently is an adjunct faculty member in the Department of Forestry and Environmental Resources at North Carolina State University, working on a National Science Foundation grant focused on association genetics in loblolly pine. He also is an affiliate faculty member at Oregon State University Department of Forest Sciences. He most recently completed a project investigating a novel approach to breeding in poplars.

DAVID WILSON

David Wilson is a professor of biochemistry and molecular and cell biology at Cornell University. He received his bachelor's degree from Harvard in 1961 and his doctorate in biochemistry from Stanford Medical School in 1966. He did postdoctoral work at the Department of Biophysics at Johns Hopkins Medical School from 1966 to 1967 before going to Cornell as an assistant professor in 1967. Wilson is a member of the American Society of Biological Chemists, American Society of Microbiologists, and American Association for the Advancement of Science. He also is a member of the Johns Hopkins Society of Scholars and director of the Cornell Institute for Comparative and Environmental Toxicology.

Wilson's laboratory uses a combination of genomics, protein engineering, and molecular biology to study the enzymology of plant cell-wall degradation, with a major focus on cellulases. Enzymes that degrade insoluble substrates have significant differences from most enzymes whose substrates are small soluble molecules. In addition, cellulases are important industrial enzymes and have potential in the production of renewable, nonpolluting fuels and chemicals. His group has been studying the high G-C gram variable soil bacterium *Thermobifida fusca*, a moderate thermophile, for more than 20 years. *T. fusca* is a major microorganism degrading plant cell walls in heated plant wastes such as compost piles. He also is using a genomic approach to compare aerobic and anaerobic organisms to discover novel mechanisms for degrading cellulose.

Research projects involve the biochemistry and chemistry of the cell walls of higher plants, with special reference to cereals and grasses and the structure-function relationships of wall polysaccharides and proteins: chemistry of lignin-carbohydrate and lignin-protein associations in walls of forage plants, with special reference to their impact on ruminant digestion; enzymology of ß-glucan hydrolases–active site chemistry in relation to specificity and to the design of specific hydrolase inhibitors; molecular mechanisms of biosynthesis of ß-glucans in plant cell walls and by bacteria; and development of monoclonal antibodies for the specific detection of cell-wall polysaccharides.

CHAPTER SOURCES

The following chapters have been previously published:

Chapter 1 to Chapter 8 – These are an edited, reformatted and augmented version of a workshop sponsored by the U.S. Department of Energy, Office of Energy Efficiency and Renewable Energy, Office of the Biomass Program, Office of Science, Office of Biological and Environmental Research Genomics: GTL Program, publication date on June 2006.

INDEX

A

abiotic, 59, 74, 82, 86, 95
absorption, 231
academic, 89, 187, 266, 288, 292, 301
acceptors, 73, 253
accessibility, 65, 132
accuracy, 161, 164, 165
acetate, 124, 177, 191, 211
acetylation, 67
achievement, xiv, 123, 191
ACI, xii, 1, 35
acid, xv, 2, 66, 71, 114, 115, 161, 172, 178, 191, 192, 194, 216, 227, 259, 285
acidic, 166
acidity, 128, 131
actinomycetes, 148, 248
activation, 107
active site, 297, 307
acute, 86
adaptability, 92
adaptation, 90, 155, 160, 163, 179, 186, 188, 195, 215
addiction, 1
additives, 268
adhesion, 152, 196, 201, 297
administration, 5
administrative, 282
adsorption, 33, 125
Advanced Energy Initiative(AEI), xii, 1, 5, 26, 35
aerobic, 152, 155, 157, 194, 233, 306
AFM, 58, 128, 231, 234
agents, 64, 147, 151
agricultural, xiii, 5, 10, 14, 15, 18, 25, 26, 69, 81, 82, 83, 84, 86, 87, 89, 100, 224, 287, 288, 289, 303
agricultural commodities, 87, 287
agricultural crop, xiii, 81, 83, 84, 100
agricultural residue, 10, 26
agriculture, xii, 4, 15, 43, 91, 95, 300
aid, 45, 58, 60, 134, 209, 218, 236, 242, 244
air, 36, 37, 168, 268
air quality, 268
alcohol, xiv, 46, 47, 50, 52, 62, 71, 136, 176, 189, 207, 210, 211,215, 216, 221
alcohol production, 215
aldehydes, 161
alkaline, 166, 301
alkanes, xiv, 176, 210, 213, 221
alleles, 92
alpha, 141, 172
alternative, xii, xiv, 1, 3, 5, 9, 24, 29, 35, 114, 142, 176, 210, 214, 217, 221, 246, 265, 269, 289
alters, 111
American Association for the Advancement of Science, 285, 306
American Competitiveness Initiative, xii, 1, 35
amino, 183, 227, 305
amino acid, 183, 227, 305
ammonia, 71
amorphous, 63
amylase, 146, 172
anaerobic, 151, 155, 194, 203, 217, 233, 290, 296, 306
anaerobic bacteria, 155, 194, 233
analytical techniques, 74, 164, 200, 228, 232
analytical tools, 93, 128, 161, 217, 228, 229
animal waste, 287
annotation, 73, 150, 158, 204, 243
annuals, 84, 85, 86, 91

anomalous, 282
antibiotics, 305
application, xi, xii, 1, 14, 20, 21, 43, 45, 48, 68, 70, 73, 93, 110, 124, 172, 176, 229, 265, 282
applied research, 43, 48, 105, 106, 210
Arabidopsis thaliana, 89, 104, 105, 120
arabinogalactan, 59
army, 151
aromatic compounds, 124
aromatic hydrocarbons, 294
aromatic rings, 137
ARS, 287
artificial intelligence, 208
ash, 98, 161
Asia, 293
asian, 293
Asian countries, 293
assault, xiii, 45, 55
assessment, 113, 148, 199, 213, 224
assets, 18
assignment, 73, 158, 243, 291
assimilation, 303
assumptions, 12, 14, 18
atmosphere, 10
atomic force, 128, 233, 235, 241
Atomic Force Microscopy (AFM), 128, 234, 235
atoms, 149, 244
ATP, 183, 190, 207, 215, 228, 253
ATPase, 297
attachment, 152
attacks, 66, 147
authority, 27
automobiles, 5
autotrophic, 303
availability, 9, 18, 93, 95, 97, 102, 112, 163, 248

B

bacteria, 20, 21, 125, 151, 153, 154, 155, 194, 199, 215, 224, 233, 242, 289, 293, 295, 297, 303, 307
bacterial, 67, 151, 158, 255, 286, 291, 292, 297, 300, 303
bacterial cells, 292
bacterial strains, 300
bacterium, 157, 306
barriers, xi, xii, xiii, 1, 2, 4, 32, 33, 45, 55, 64, 65, 71, 99, 131, 145, 149, 175, 176, 200, 222, 244, 245
base pair, 114
basic research, 72, 106, 116
battery, 281

behavior, 31, 159, 234, 256, 284, 303
benchmark, 166
beneficial effect, 69
benefits, 7, 10, 11, 27, 191, 192, 239
benign, 23, 45, 68, 72
beverages, 173
binding, 46, 74, 125, 128, 138, 144, 147, 152, 185, 189, 207, 235, 253, 284, 296, 297
biocatalysis, xiv, 171, 172, 219, 288
biocatalysts, 22, 52, 167, 171, 173, 175, 176, 177, 192, 194, 202, 214, 215, 216, 217, 218, 223
biocatalytic process, 22
biochemistry, 57, 145, 148, 155, 281, 289, 291, 293, 296, 298, 299, 300, 301, 302, 303, 306
bioconversion, 27, 46, 64, 67, 70, 128, 133, 171, 172, 173, 175, 202, 210, 222, 223, 225, 233, 237, 244, 253, 268, 287, 288, 290, 293
biodegradable, 289, 294
biodegradation, 63, 150, 193, 298
biodiesel, xi, 2, 114, 115, 116
bioengineering, 294, 295, 299, 302
bioethanol, 9, 62, 82, 89, 99, 142, 155, 217, 218
biofilms, 237, 238
biogenesis, 294
bioinformatics, 93, 173, 296, 302, 305
biological interactions, 256
biological models, 244
biological processes, 4, 24, 89, 235
biological systems, 14, 29, 31, 32, 127, 210, 284, 294
biomarkers, 73, 112, 131, 164, 249, 262
biomass materials, xv, 259, 261, 262
biomaterials, 297
biomimetic, 291
biomolecular, 150
biomolecules, 301
biophysics, 68, 281, 283, 301
biopolymers, 136, 137, 162, 163, 164, 183, 284, 297
bioreactor, 152, 183, 238, 246, 289
biorefinery, xi, xii, 6, 12, 14, 15, 22, 24, 26, 27, 43, 45, 48, 53, 56, 57, 69, 124, 127, 133, 146, 148, 149, 174, 186, 222, 237, 246, 260, 262, 266, 267
biosensors, 295
biosorption, 288
biosphere, 46, 146, 149, 289, 303
biosynthesis, 61, 62, 72, 73, 75, 90, 104, 109, 110, 114, 183, 243, 245, 248, 250, 285, 290, 294, 298, 303, 307
biosynthetic pathways, 190, 286
biotechnological, 15, 105, 296

biotechnology, xii, 4, 14, 29, 30, 31, 43, 196, 200, 201, 203, 210, 266, 284, 286, 288, 291, 293, 295, 296, 297, 300, 301, 302, 304
biotic, 59, 74, 82
biotin, 283
biotransformation, 205, 208, 209, 245
birds, 85
bleaching, 282, 297, 301
bonding, 66
bonds, 145
bottlenecks, 67, 133, 198, 199, 244
branching, 141
Brazil, 7
breakdown, xiii, xiv, 20, 21, 27, 44, 46, 51, 63, 64, 67, 81, 123, 124, 132, 134, 152, 196, 225, 226, 248, 303
breeder, 306
breeding, 14, 67, 72, 74, 81, 82, 92, 93, 94, 96, 98, 99, 109, 163, 164, 224, 229, 249, 305, 306
building blocks, 34, 59, 244
Bureau of Economic Analysis, 37
burning, 9
burns, 114
bypass, 162
by-products, 18, 34, 167, 172, 181, 191, 206, 215, 269

C

CAD, 71
caffeic acid, 71
calcium, 61
calibration, 160, 164
candidates, 133, 150
capillary, 160, 227
capital cost, 146, 165
capital expenditure, 176, 213
carbohydrate, 23, 40, 65, 68, 73, 94, 113, 114, 115, 117, 128, 130, 132, 144, 145, 151, 160, 164, 205, 206, 208, 216, 223, 226, 229,230, 244, 252, 256, 260, 283, 284, 295, 297, 303, 306
carbohydrate metabolism, 164, 206, 303
carbon cycling, 134, 303
carbon dioxide, 3, 7, 9, 198, 303
carbon monoxide, 11
carboxylic acids, 161
carcinogenic, 114
catalysis, 147, 217, 290
catalytic activity, 110, 132, 140, 245
catalytic properties, 139

causality, 254
CCR, 71
cDNA, 107, 304
cell adhesion, 152
cell division, 94, 211
cell growth, 181, 185, 206, 256
cell line, 286
cell metabolism, 237
cell surface, 152, 234, 251
cellulosic, xi, xii, xiii, 1, 2, 3, 5, 6, 7, 9, 10, 11, 12, 13, 19, 21, 22, 24, 27, 35, 40, 65, 123, 124, 127, 171, 176, 194, 198, 222, 223, 232, 263, 267, 268, 269, 282, 293, 295
cellulosic ethanol, 2, 4, 5, 6, 7, 10, 11, 12, 28, 124, 172, 268, 269, 293
cellulosomes, 33, 52, 140, 142, 143, 144, 147, 148, 149, 151, 153, 154, 233, 249, 250, 283
census, 37
census bureau, 37
CEO, 281, 305
cereals, 303, 306
certificate, 285
chemical bonds, 128, 223, 229
chemical degradation, xi
chemical energy, 199
chemical engineering, 282, 287, 288, 290, 291, 294, 295, 302
chemical pretreatments, 66
chemical properties, 127
chemicals, 5, 26, 27, 34, 57, 58, 65, 112, 130, 145, 154, 163, 190, 210, 228, 266, 268, 285, 287, 288, 300, 302, 306
chloroplast, 294
chromatograms, 74
chromatography, 73, 227
chromosomes, 93
classes, 46, 136, 232, 286
classical, 148, 205
Clean Air Act, 267, 268, 269
clean-energy, 5
cleavage, 133, 141, 142
climate change, 3, 7
cloning, 149, 286
clusters, 243
CO_2, 3, 9, 10, 69, 86, 91, 94, 95, 198, 199, 213, 303
cobalt, 282
codes, 129, 149, 244
cofactors, 150, 156, 252
cohesins, 144, 148, 152, 153, 154
collaboration, 33, 239, 282

colonization, 237
combustion, 114, 217, 266
commercialization, 5, 29, 188, 295, 304
commodity, 172, 173, 224
communication, 239, 242, 292
compatibility, 139
competitive advantage, 152
competitive process, 177
complement, 46, 73, 115, 144, 191, 195, 205
complex interactions, 104, 150
complex systems, 94, 165
complexity, 22, 32, 34, 125, 160, 165, 171, 182, 191, 192, 194, 203, 205, 241, 243, 246, 255
composites, 130, 282, 297
compost, 306
compounds, xiii, 6, 56, 57, 75, 114, 115, 124, 164, 167, 178, 183, 185, 194, 199, 210, 211, 212, 213, 217, 228, 229, 282, 285, 289, 291, 298
computation, 250, 255, 299
computational modeling, 4, 52, 144
computer architecture, 241
computer graphics, 284
computer science, 57
computing, 29, 32, 185, 205, 209, 247, 255, 296, 304
concentration, 156, 176, 195, 197, 198, 210, 228, 229, 292
concrete, 21
conducting polymers, 282
configuration, 137, 192
congress, vi, 6, 28, 37, 265, 270
conjugation, 283
consensus, 161, 162, 200
conservation, 9, 265
consolidation, xiv, 51, 56, 123, 127, 191, 194, 195, 197, 198, 201, 202, 222, 223
constraints, 147
construction, 16, 144, 164, 214, 226, 235, 252, 283, 285
construction and demolition, 16
consulting, 282, 284
consumers, 177
consumption, 6, 8, 9, 10, 13, 14, 15, 40, 179, 181, 229
contamination, 197, 203, 234
content analysis, 228
contracts, 103
control, 34, 46, 47, 49, 50, 53, 59, 69, 70, 74, 75, 89, 92, 95, 99, 108, 110, 112, 126, 136, 137, 147, 154, 158, 175, 179, 183, 188, 190, 191, 195, 196, 197, 204, 205, 227, 245, 248, 250, 289, 294, 298, 303

conversion rate, 83, 172
cooling, 25
copper, 288
corn, xiv, xv, 2, 4, 7, 10, 11, 12, 14, 15, 18, 20, 26, 40, 41, 44, 56, 57, 74, 85, 95, 96, 100, 101, 103, 106, 123, 126, 146, 171, 172, 214, 259, 282, 304
correlation, 111, 125, 161, 231, 260, 261, 295
corrosive, 213
cost benefits, 192
cost-effective, xiii, 56, 81, 87, 111, 195, 202, 248, 260, 266, 289
costs, xiii, xiv, 4, 18, 25, 27, 50, 56, 69, 82, 84, 90, 92, 115, 116, 145, 146, 157, 165, 171, 172, 191, 210, 213, 214, 216, 249, 270
coupling, 136, 261
covalent, 63, 64, 66, 135
covalent bond, 135
crop residues, xiv, 17, 18, 44, 83, 100, 102, 123, 159
croplands, 15, 18, 104
cross-linking, 59, 63, 298
crude oil, 15
crystal lattice, 146
crystal structure, 282
crystalline, 45, 63, 125, 134, 142, 145, 151, 157
crystallinity, 63, 66, 137
crystals, 66, 128
cues, 252
cultivation, 82, 200, 204, 223, 230, 237, 238, 256, 288
cultivation conditions, 230
cultural practices, 100
culture, xiv, 22, 23, 27, 34, 123, 150, 175, 194, 200, 201, 203, 204, 237, 238, 253
culture conditions, 150, 237, 253
current limit, 215
cuticle, 65
cycling, 100, 134, 303
cytochrome, 62
cytoskeleton, 64, 110
cytosol, 110, 250

D

dairy, 287
data mining, 154
data set, 157, 160, 206, 207, 208, 245, 246, 254
database, 106, 298
data-mining, 239
de novo, 205, 208
decay, 157, 158

decision making, 242
decisions, 242
decoding, 113, 115
decomposition, 100, 139, 222, 223, 248, 255
deconstruction, xiv, xv, 32, 44, 45, 46, 49, 51, 53, 65, 70, 123, 124, 129, 131, 133, 155, 158, 194, 222, 224, 232, 233, 235, 248, 249, 250, 253, 259
decoupling, 181
defense, 285
Defense Advanced Research Projects Agency, 297
deficits, 15
definition, 93, 243
degradation, xi, 21, 33, 63, 71, 75, 134, 141, 143, 147, 148, 150, 152, 154, 155, 156, 157, 164, 165, 200, 201, 206, 224, 226, 228, 243, 244, 256, 260, 262, 283, 294, 297, 306
degrading, 46, 133, 143, 144, 151, 154, 166, 226, 232, 244, 262, 286, 306
dehydration, 73, 112, 127, 128, 129, 213
dehydrogenase, 71, 189
delivery, 25, 294
demonstration project, 266
denaturation, 66
density, 58, 74, 106, 116, 213, 304
Department of Agriculture, 5, 36, 37, 82, 113, 117, 273, 274, 275, 276, 277, 278, 280, 282, 287, 293, 298
Department of Commerce, 293
Department of Energy (DOE), ix, 1, 2, 25, 36, 37, 38, 43, 55, 81, 82, 123, 171, 221, 231, 257, 259, 265, 271, 273, 275, 276, 277, 279, 280, 281, 293, 298, 300, 309
depolymerization, 72, 124, 133, 134, 141, 226, 260, 303
deposition, 59, 61, 62, 69, 74, 115
depressed, 10
derivatives, 136, 299
detection, 156, 163, 227, 231, 249, 295, 307
detergents, 145
detoxification, 22, 165, 167, 191, 192, 193, 195, 197
developing countries, 9
developmental process, 114
diabetes, 302
diamonds, 20
diesel, xii, 1, 5, 7, 25, 40, 70, 116, 213, 266, 267
diesel fuel, xii, 1, 5
differentiation, 286
diffraction, 128, 231
diffusion, 208, 231, 245
digestibility, 57, 66, 99, 131

digestion, 127, 143, 203, 232, 307
dimensionality, 32
dimeric, 137
discipline, xi, xii, 43
discovery, 104, 105, 118, 216
displacement, 7
disseminate, 163, 239, 247
distillation, 214
distribution, 8, 9, 209, 230, 242, 269
diversity, 45, 46, 50, 74, 85, 103, 133, 134, 135, 141, 143, 146, 149, 151, 155, 158, 190, 195, 196, 214, 224, 249, 266
division, 94, 211
DNA, 30, 68, 70, 87, 93, 94, 103, 105, 106, 107, 111, 112, 113, 115, 120, 156, 159, 160, 162, 181, 207, 211, 223, 224, 230, 240, 248, 281, 296, 301
DNA repair, 281
DNA sequencing, 87, 93, 112, 160, 248, 301
dockerins, 144, 148, 152, 153
domestic resources, 4, 25
domestication, 52, 92, 99, 304
donor, 216, 303
draft, 304
drought, 52, 86, 93, 95, 96, 97, 106, 107
drug design, 284
drugs, 294
DuPont, 186, 274, 292
durability, 292
duration, 95

E

E. coli, 134, 177, 178, 180, 183, 185, 186, 293, 294, 305
earth, 4, 134, 303
ecological, 103
ecology, 111, 287, 290, 303
economic competitiveness, 29
economic development, 7, 267
economic growth, 3, 7, 24
economic policy, 26
Economic Research Service, 37
economic security, 25, 28
economics, 139
economies of scale, 31
ecosystems, 31, 44, 49, 99, 101, 102, 134, 225, 236, 243, 292
education, 275, 276, 277
effluent, 282
election, 294

electricity, 11, 12, 23, 25, 266, 270, 285
electrochemistry, 285
electron, 73, 95, 128, 136, 216, 231, 233, 234, 235, 241, 289, 303
electron microscopy, 128, 234, 235
electronic structure, 287
electrophoresis, 160, 227
encapsulation, 292
encoding, 62, 94, 134, 144, 166, 181
endoplasmic reticulum, 111
endosperm, 116
end-to-end, 53
energy efficiency, 7, 25, 265
Energy Efficiency and Renewable Energy (EERE), 2, 25, 27, 28, 37, 86, 117, 271, 273, 275, 276, 277, 285, 309
energy emission, 3
Energy Information Administration, 36
Energy Policy Act, ix, 5, 37, 265, 270
Energy Policy Act of 2005, ix, 5, 37, 265, 270
energy supply, 6
energy transfer, 110
engagement, 235
engines, 114, 266
enterprise, 10
environment, 11, 24, 28, 31, 32, 114, 161, 175, 178, 179, 186, 204, 222, 227, 234, 237, 239, 247, 250, 256, 257, 304
environmental conditions, 147, 200, 223, 225, 226
environmental contaminants, 294
environmental influences, 304
Environmental Protection Agency (EPA), 269
enzymatic, 20, 21, 22, 23, 45, 46, 56, 57, 59, 65, 66, 67, 68, 69, 127, 129, 132, 133, 136, 139, 141, 142, 146, 152, 153, 156, 167, 181, 183, 194, 196, 198, 229, 230, 232, 241, 244, 260, 284, 296, 301
enzyme immobilization, 289
enzyme interaction, 45, 56
epidermis, 65
epitope, 70, 110
erosion, 11, 100
Escherichia coli, 178, 186, 219
EST, 92, 93, 107, 108, 304
esterases, 135, 141, 151, 241, 250
esters, 115, 161, 210, 211, 262
estimating, 14, 230
ethers, 136
ethylene, 305
European Commission, 103

evolution, 89, 139, 140, 148, 150, 179, 185, 187, 195, 201, 203, 209, 212, 222, 225, 243, 248, 262, 293
excision, 281
excitation, 231
excretion, 150, 227
expansions, 20
expertise, 2, 290, 292, 297, 298, 304, 305
exposure, 86, 212
expressed sequence tag, 92, 224
extracellular matrix, 242
extraction, 165, 282
extremophiles, 215

F

family, 89, 105, 143, 147, 232, 243
family members, 143
Farm Bill, 6
Farm Security and Rural Investment Act, 37
farmers, 10
farming, 25, 100
farmland, 96
farms, 254
fatty acids, xiv, 116, 210, 221, 230, 299
feedback, xv, 246, 259
feeding, 230
feedstock, xiii, xv, 3, 5, 6, 9, 25, 44, 53, 60, 65, 69, 70, 73, 82, 89, 93, 109, 112, 113, 123, 133, 156, 157, 160, 161, 162, 166, 171, 224, 229, 232, 248, 255, 259, 266, 267, 291
feet, 13
fermentation broth, 214, 215
fertility, 19, 83, 92, 99, 101, 105, 107
fertilization, 82
fertilizer, 11, 74, 84, 85, 92, 197
ferulic acid, 161, 285
fiber, xiii, 2, 14, 15, 18, 19, 20, 65. 81, 89, 102, 128, 145, 293, 296, 297, 305
fibrils, 110, 124, 142
fidelity, 129, 137
field trials, 129
filtration, 246, 300
financing, 199
fingerprinting, 73
fire, 16, 270
fire hazard, 16
first generation, 183
fixation, 69, 86, 91, 94, 95, 303
flex, 21

flexibility, 202
flow, 7, 75, 92, 290
fluorescence, 110, 301
folding, 297
food, xiii, 14, 15, 16, 18, 19, 20, 81, 82, 89, 96, 102, 104, 116, 203, 288, 295, 300
food industry, 203
food production, 96, 104
food products, 20
forest management, 304
Forest Service, 282, 283
forestry, 5, 14, 15, 83, 120, 278, 293, 298, 299, 304, 306
fossil, 3, 9, 10, 11, 12, 13, 266, 289, 294
fossil fuel, 9, 10, 11, 12, 13, 266, 289, 294
fourier, 74, 109, 128
fourier transform infrared spectroscopy, 74, 109
fractionation, 285, 290, 291
freezing, 86
fructose, 20, 173
FTIR, 109, 128, 160
fuel, xiii, xiv, 4, 5, 6, 7, 9, 10, 11, 12, 14, 15, 20, 21, 27, 40, 41, 43, 53, 55, 57, 68, 69, 70, 86, 114, 171, 211, 213, 217, 228, 229, 230, 256, 266, 267, 268, 269, 285, 291, 293
fuel cell, 266, 285
functional analysis, 155
funding, 7
funds, 6, 10, 268
fungal, 143, 151, 196, 292
fungi, 63, 101, 125, 132, 134, 148, 150, 194, 224, 242, 248, 253, 292
fungus, 100, 134, 292
furniture, 16
fusion, xii, 4, 43

G

gas, 12, 73, 217, 218, 227
gas chromatograph, 73, 227
gasification, 218
gasoline, xii, 1, 3, 4, 5, 6, 7, 9, 10, 11, 13, 25, 40, 124, 266, 267, 268
gastrointestinal, 287, 297
GEC, 6, 7, 8, 36
gel, 61
gelation, 67
gene amplification, 181
gene expression, 92, 99, 211, 212, 227, 238, 245, 282

gene silencing, 305
gene transfer, 151
General Motors, 36
generation, xii, 1, 3, 4, 29, 31, 44, 47, 51, 57, 104, 105, 112, 127, 160, 187, 209, 225, 231, 232, 248, 281, 286, 299, 304
genetic blueprint, 224
genetic control, 95, 99, 116, 183
genetic diversity, 134, 158
genetic information, 109
genetically modified organisms, 262
genetics, 92, 98, 105, 124, 152, 155, 282, 301, 304, 305, 306
genome sequences, 30, 92, 108, 114, 154, 157, 211, 216, 243, 247, 248, 249
genomic, 23, 30, 31, 32, 45, 66, 68, 87, 93, 95, 101, 102, 106, 114, 123, 124, 130, 132, 134, 143, 149, 156, 158, 160, 162, 163, 165, 186, 187, 242, 246, 247, 250, 292, 297, 306
genotypes, 92, 99, 107, 108, 109, 186, 291
geothermal, 25
GHG, 10
global demand, 15
glucoamylase, 172
glucose, xiv, 20, 21, 46, 60, 63, 126, 133, 145, 171, 173, 177, 180, 181, 182, 185, 191, 305
glucosidases, 145, 261
glycans, 253
glycerol, 181
glycolysis, 175, 180
glycoside, 135, 149, 155, 301
glycosyl, 125, 145, 248, 253
goals, xiii, 3, 24, 27, 31, 32, 33, 43, 44, 57, 82, 83, 86, 93, 99, 106, 135, 146, 160, 164, 172, 180, 197, 203, 209, 242, 255, 289, 292
government, vi, 9, 10, 270, 285, 286, 293, 306
graduate students, 304
grain, 4, 11, 18, 20, 40, 41, 100, 146, 173, 282, 289, 293, 303
grains, 11, 20
Gram-negative, 157
Gram-positive, 201
grants, 6, 269
granules, 172
graph, 186
grasses, 11, 18, 65, 82, 83, 85, 98, 104, 105, 107, 159, 160, 224, 262, 303, 305, 306
grasslands, 15
green fluorescent protein, 238
Green Revolution, 82

greenhouse, 3, 9, 36, 38, 102, 119, 120, 289, 296, 303
greenhouse gas (GHG), 3, 9, 102, 289, 296, 303
groups, 27, 35, 67, 69, 161, 162, 188, 227, 229, 246, 260, 283, 286
growth rate, 95, 185, 196
guidance, 202
guidelines, 104
gums, 161
gut, 202

H

habitat, 85
habitation, 11
handling, 222, 246, 287
hands, 165
hanging, 141
hardwoods, 57
harmful effects, 72, 289
harvest, xiii, 5, 41, 66, 81, 84, 90, 103, 115, 158, 159, 166, 214, 262
harvesting, 14, 18, 66, 69, 84, 90, 99, 101
hazards, 16
health, 16, 83, 303
heart, 240
heat, 11, 23, 146, 214, 246, 266, 270
heating, 16, 25, 128
heavy metal, 294
hematopoietic, 286
hemicellulose, 21, 33, 40, 56, 61, 63, 64, 65, 67, 72, 94, 125, 131, 133, 134, 135, 136, 139, 140, 151, 163, 166, 193, 216, 232, 248, 260, 262, 297
hemicellulose hydrolysis, 193
herbicide, 11, 292
herbivores, 297
heterogeneity, 66, 129, 227
heterogeneous, 125, 143, 153, 226, 237, 238
heterogeneous systems, 226
heterosis, 99
high resolution, 73
high temperature, 172, 217, 218
high-level, 33, 35, 116, 150
high-performance liquid chromatography, 216, 227, 291
high-throughput screening, 92, 99, 124
high-value products, 292
hips, 5, 70, 102, 111, 211, 248
histochemical, 62
holistic, 92, 93, 106, 149, 154

holistic approach, 92
hormones, 304
horticulture, 280
host, 110, 134, 139, 144, 147, 148, 149, 175, 180, 183, 189, 191, 193, 194, 200, 201
human, 114, 173, 175, 267, 288, 296, 303
human genome, 114
Human Genome Project, 304
human subjects, 267
hybrid, 11, 56, 99, 105, 144, 147, 148, 281
hybridization, 238
hydration, 284
hydrides, 282
hydro, 146, 294
hydrocarbon, 193
hydrogen, 60, 66, 145, 285, 289
hydrolases, 125, 135, 145, 149, 155, 216, 301, 307
hydrolysates, 71, 177, 178, 188, 191, 198, 216
hydrolyzed, 126, 198
hydrophobic, 61, 64, 210, 211, 213
hydropower, 25
hydrothermal, 45
hydroxyl, 161
hydroxyl groups, 161
hydroxylation, 62
hypothesis, 137, 157, 187, 298

I

identification, 67, 89, 93, 98, 104, 108, 110, 111, 132, 139, 140, 162, 189, 190, 200, 203, 211, 212, 224, 230, 237, 248, 250, 251, 289
identity, 63, 115, 116, 210
IEA, 6, 36, 167
image analysis, 57, 109
images, 57, 72, 112, 160, 163, 234, 235
imaging, 45, 47, 57, 72, 73, 110, 111, 112, 113, 124, 126, 128, 129, 141, 142, 149, 156, 162, 173, 204, 222, 223, 231, 233, 235, 236, 238, 240, 241, 242, 246, 249, 251, 256, 257, 261, 262, 297
imaging modalities, 72
imaging techniques, 126, 238, 256
immobilization, 233, 288
implementation, 11, 48, 93, 175, 239, 246
imports, 3, 8, 9
n situ, 72, 112, 231, 238
in situ hybridization, 238
in vitro, 67, 110, 154, 176, 212, 249
in vivo, 67, 133, 190, 206, 207, 228, 242, 249, 286, 299

inactive, 137
incentives, 6, 265, 267
inclusion, 207
income, 10, 199
incompatibility, 105
Indian, 270
industrial application, 139
industrial experience, 300
industrial processing, xi, xii, 4, 43
industrial production, 187, 202
industry, xi, xii, xiii, xiv, 2, 4, 5, 9, 10, 11, 12, 15, 17, 27, 29, 31, 35, 43, 57, 81, 108, 109, 146, 148, 171, 172, 175, 176, 192, 203, 266, 267, 269, 282, 285, 287, 288, 291, 300, 306
information system, 251, 272, 276
infrared, 128, 160, 231, 291
infrared spectroscopy, 74, 109, 160, 291
infrastructure, 4, 5, 7, 27, 30, 31, 90, 92, 99, 213, 222, 237, 239, 240, 242, 265, 286, 305
inhibition, 33, 50, 175, 176, 191
inhibitors, 22, 33, 34, 45, 47, 49, 50, 52, 56, 124, 127, 159, 161, 178, 179, 184, 185, 191, 197, 199, 222, 229, 307
inhibitory, xiii, 56, 75, 124, 167, 177, 178, 185, 188, 190, 212
initiation, 75, 98, 286
injury, vi
Innovation, 35, 37, 292
inorganic, 161, 198, 218, 282
inorganic salts, 161
insertion, 105, 108, 110, 162
insight, 146, 154, 159, 186, 225, 229, 254, 262
institutions, 26, 93, 242, 266, 286, 301
instruments, 159, 160, 161, 162, 231, 254
integration, 26, 90, 161, 242, 243, 260, 261, 287, 291, 300
intelligence, 293
interdisciplinary, 68, 286, 287, 288
interface, 161, 195, 233, 244, 251, 287
intermolecular, 287
intermolecular interactions, 287
internal combustion, 266
International Energy Agency, 36
International Trade, 37
interrelationships, xv, 129, 221
intrinsic, 202
invasive, 98
Investigations, 31, 32, 247
investigative, 242
investment, 3, 30, 73, 81, 82, 160, 172

ionization, 227
ions, 287
iron, 282
isolation, 66, 156, 164, 209, 215, 229, 251, 291
isotopes, 183, 190, 230
isotopic methods, 302

J

jobs, 9
Jung, 61, 121, 167

K

kernel, 2, 116
ketones, 161
kinetics, 125, 207, 245, 296, 298
king, 298
knockout, 181

L

labeling, 73, 190, 204, 230, 256
labor, 92, 116, 162
lactic acid, 2, 194
Lactobacillus, 215
land, 13, 14, 15, 18, 27, 85, 97, 99, 103, 197, 270
land use, 27
large-scale, 7, 8, 13, 14, 15, 18, 29, 56, 82, 88, 101, 115, 135, 145, 165, 185, 246, 296
laser, 238
law, 265
leadership, 129, 285, 287, 296
learning, 142, 166
legume, 114, 292
liberal, 283
life cycle, 10, 73, 105, 106, 112, 213, 233
life sciences, 288
life-cycle cost, 213
ligand, 283
lignans, 161
lignocellulose, 46, 50, 128, 144, 154, 155, 156, 157, 166, 173, 194, 214, 216, 224, 225, 243, 248, 260
limitations, 18, 95, 130, 139, 140, 148, 162, 171, 172, 176, 183, 200, 208, 210, 222, 233, 245
linear, 60
linkage, 64
links, 32, 64, 166, 236
lipases, 301

lipids, 115, 116, 117, 213, 215, 216, 223, 252, 299, 300, 301, 303
liquid chromatography, 164, 216, 227, 291
liquid fuels, xi, xii, xiii, xiv, 43, 44, 46, 70, 88, 98, 115, 171, 221, 223, 224, 237, 246, 255, 260
liquor, 16
lithium, 281
livestock, 20, 140
loan guarantees, 268, 269
local government, 9, 270
localization, 70, 110, 208, 246, 249, 250, 251
locus, 98
long period, 100
long-term impact, 19
low molecular weight, 228
lysine, 183

M

machinery, 25, 205, 241, 244
machines, 50, 52, 110, 142, 146, 147, 148, 218, 226, 231, 242, 244, 245, 249, 250, 251, 256, 289
macromolecules, 56, 127, 227, 230, 298
magnetic, vi, 73, 113, 128, 190, 228, 282, 298
magnetic resonance, 73
magnetic resonance imaging, 73
maintenance, 45, 115, 207, 215
maize, 89, 114, 117
mammalian cell, 302
management, 18, 99, 102, 103, 104, 214, 234, 239, 247, 250, 282, 283, 286, 290, 304, 305
management practices, 100, 103, 305
mandates, 5, 6
manganese, 151, 298
manipulation, 34, 49, 50, 52, 53, 113, 134, 175, 180, 184, 189, 194, 198, 230, 250, 256, 301
manufacturing, 266, 289, 293, 295
manure, 17, 18
mapping, 74, 187, 190, 208, 224, 231, 304
market, xiii, 2, 9, 27, 56, 213
marketing, 293
marketplace, 29, 30
mass spectrometry, 73
mass transfer, 218
mathematical methods, 129
mathematicians, 304
mathematics, 241
matrix, 45, 110, 134, 142, 160, 165, 242
maturation, 73, 112, 233
measurement, 203

measures, 89, 244
mechanical properties, 299
media, 178, 186, 188, 191, 197, 198, 217
Medicago truncatula, 292
medium composition, 218
melanoma, 286
membranes, 189, 211, 212
mesoscopic, 130
metabolic intermediates, 303
metabolic pathways, 64, 90, 110, 124, 176, 202, 206, 208, 230, 245, 250
metabolic systems, 53, 207
metabolism, 47, 52, 69, 104, 142, 164, 176, 177, 179, 181, 187, 190, 193, 196, 206, 208, 215, 216, 222, 226, 227, 228, 230, 237, 244, 245, 255, 256, 285, 290, 294, 299, 302, 303
metabolites, 134, 181, 183, 190, 196, 199, 207, 227, 228, 230, 231, 232, 235, 236, 242, 252, 253, 254, 285
metabolizing, 46
metabolome, 201, 203
metabolomics, 45, 92, 159, 173, 185, 228, 229, 252, 294
metagenomics, 154
metallic catalysts, 291
metals, 218, 286
metric, 3, 5
microarray, 120, 226
microbial cells, 152, 230, 236, 237, 245, 251, 263
microbial communities, 19, 24, 33, 44, 83, 100, 101, 102, 103, 104, 111, 154, 155, 156, 157, 158, 159, 176, 179, 202, 203, 204, 221, 223, 224, 225, 238, 241, 247, 255, 256, 296
microchip, 150
microflora, 169
microorganisms, xiv, 4, 34, 45, 47, 57, 104, 148, 150, 151, 171, 173, 175, 177, 184, 196, 198, 199, 200, 202, 205, 210, 211, 214, 215, 218, 221, 222, 225, 226, 230, 246, 247, 255, 260, 263, 287, 294, 295, 296, 303, 306
microRNAs, 248
microscopy, 73, 110, 128, 231, 234, 235, 238, 287
microtubules, 64
milligrams, 160
mineralization, 294
minerals, 178
mining, 45, 154, 190, 239, 288
missions, 3, 7, 28, 29, 267
mitochondria, 207
mixing, 5, 125

modalities, 72
model system, 33, 104, 107, 113, 285
modulation, 183
modules, 43, 49, 144, 152, 153, 154, 164
moieties, 127, 128, 137, 223, 256
moisture, 15, 59, 234
molecular biology, 57, 124, 173, 260, 262, 286, 291, 296, 297, 305, 306
molecular dynamics, 244, 284, 287
molecular markers, 92, 93, 107
molecular mechanisms, 126, 179, 222, 307
molecular structure, 65, 233
molecular weight, 228
molecules, 60, 63, 67, 127, 142, 145, 149, 152, 164, 204, 210, 228, 229, 234, 235, 236, 238, 252, 254, 256, 268, 282, 287, 306
monoclonal, 307
monoclonal antibodies, 307
monomers, 20, 61, 62, 72, 75, 140, 285, 305
monosaccharides, 133, 262, 263
morphological, 296
mRNA, 70, 111
MSW, 16, 18
multidisciplinary, 260, 290, 297
multiphase flow, 290
multiplexing, 160
multivariate, 160, 291
municipal solid waste, 16
muscle, 297
mutagenesis, 248, 284
mutants, 62, 107, 108, 110, 140, 144, 147, 183, 185
mutations, 105, 147, 150, 225, 284
MVA, 160, 162, 164
mycology, 302

N

NAD, 190, 207, 253
nanobiotechnology, 283
nanometer, 149, 231, 233
nanometer scale, 149, 231, 233
nanoscale materials, 212
nanoscience, 149
nanostructured materials, 282
nanotechnology, 297
nation, xiii, 2, 5, 6, 14, 15, 27, 30, 43, 44, 213, 296
National Academy of Sciences, 62, 293, 301
National Economic Council, 35
National Institute of Standards and Technology (NIST), 167, 282

National Institutes of Health (NIH), 281, 296, 301
National Research Council, 288, 295, 302
National Science Foundation, 274, 275, 296, 298, 300, 306
national security, 9, 267
natural, xiii, 12, 45, 50, 55, 57, 66, 93, 94, 109, 134, 143, 149, 151, 154, 155, 156, 158, 164, 165, 196, 203, 204, 230, 234, 248, 249, 256, 286, 288, 293, 294, 296
natural adaptations, 151
natural gas, 12
natural resources, 288
natural science, 286
natural selection, 196, 203
near infrared spectroscopy, 160
nerve, 294
nerve agents, 294
network, 4, 60, 208, 226, 230, 239
neutralization, 191, 192
new frontier, 117
next generation, 160
niche market, 57
nitrogen, 69, 85
NMR, 73, 75, 113, 128, 141, 190, 207, 228, 230, 233, 253, 254, 256, 282, 298
non-native, 97, 136, 191
nontoxic, 210
normal, 72, 94
North America, 36, 119, 297, 301
northeast, 82, 83
nuclear, 73, 111, 113, 128, 190, 228, 282, 294, 298
nuclear magnetic resonance, 73, 113, 128, 190, 228, 282, 298
nucleic acid, 227, 249
nucleotide sequence, 112
nucleotides, 110, 250
nutrient, 11, 15, 19, 44, 49, 69, 82, 83, 84, 92, 100, 102, 104, 108, 114, 115, 178, 197, 198, 207, 214, 292
nutrient cycling, 100
nutrition, 300, 303

O

obesity, 302
observations, 137
octane, 40
Office of Science and Technology Policy, 35
oil, xii, 1, 3, 5, 7, 8, 9, 15, 20, 27, 35, 70, 114, 115, 116, 117, 299

oil palm, 115
oil production, 3, 116
oilseed, 114, 117
oligomeric, 229
oligomeric products, 229
oligomers, 298
oligosaccharides, 56, 141
optical, 137, 224, 231, 233, 235, 241, 287
optical activity, 137
optics, 234
optimization, 44, 59, 127, 130, 188, 212, 253, 282, 291, 305
organ, 291
organelle, 228
organic, 16, 51, 70, 72, 82, 83, 100, 101, 164, 178, 210, 230, 285, 286, 288, 291, 293, 298, 305
organic compounds, 288
organic food, 16
organism, xiv, 22, 23, 30, 46, 71, 105, 107, 108, 134, 157, 179, 180, 183, 184, 185, 186, 189, 195, 196, 200, 210, 221, 230, 238, 243, 249, 252, 256, 293
orientation, 64, 110, 282
oxidation, 291
oxygen, 145, 289

P

Pacific, 239, 252, 271, 273, 274, 277, 281, 287, 290, 296
palm oil, 116
parallel computation, 75
parenchyma, 58, 111, 253
parentage, 92
parents, 98
partnerships, 285
pasture, 15, 18, 83
patents, 282, 285, 286, 288, 291, 293, 295, 296, 298, 300, 301, 302
pathogenesis, 303
pathogenic, 295
pathogens, 86, 98
pathology, 82, 305
pattern recognition, 163
pectins, 59, 67, 161
peer, 285, 287, 290, 291, 293, 296
peptide, 227, 305
performers, 156
periodicity, 66
permeability, 211
permeation, 268

permit, 53, 59, 66, 70, 124, 142, 143, 234, 249
perturbation, 199
pesticides, 294
pests, 86, 98
petroleum, 3, 5, 7, 24, 81, 114, 177, 266, 288
petroleum products, 3
pharmaceutical, 145, 172, 183, 185, 203, 305
pharmaceutical industry, 172
phase transitions, 282
phenolic, 161, 166
phenomenology, 31
phenotypes, 92, 162, 175, 183, 186, 201, 205, 250, 291
phenotypic, 92, 200
phenylalanine, 71
phloem, 65, 111, 253
phosphate, 292, 294
phosphorus, 292
phosphorylation, 111
photodegradation, 282
photosynthesis, 75, 85, 95, 114
photosynthetic, 9, 10, 69, 86, 91, 94, 95, 116, 211, 289, 303
photosynthetic systems, 211, 289
physical properties, 73, 246, 297
physical sciences, 29
physicochemical, 66, 102, 130, 156, 208, 245, 256
physics, 149, 282, 287
physiological, 179, 185, 200, 237, 238, 256, 306
physiology, 102, 175, 183, 195, 208, 209, 227, 228, 294, 296, 297, 299, 302
pilots, 247
pipelines, 164
pith, 58
planning, 6, 233, 281
plasma, 64, 110, 297
plasma membrane, 64, 110
plastic, 145, 268, 285
platforms, 34, 164, 175
play, 29, 93, 100, 105, 112, 151, 189, 199, 229, 248, 252
PLC, 271, 281
Pleurotus ostreatus, 168
point mutation, 284
pollutants, 114, 286, 289
polyester, 305
polymeric materials, 299
polymerization, 61, 63, 75, 137, 164, 298
polymerization mechanism, 137

polymers, xi, 19, 21, 23, 27, 44, 45, 57, 59, 63, 64, 67, 68, 69, 71, 72, 75, 94, 95, 127, 128, 131, 133, 134, 136, 140, 145, 146, 172, 223, 226, 228, 229, 233, 249, 282, 284, 287, 289, 294, 297
polymorphism, 92, 93, 94, 106
polypeptide, 138
polysaccharides, 63, 64, 76, 169
poor, 111
population, 19, 83, 153, 154, 200, 203, 226, 238, 282
pores, 126
portfolio, 3, 6, 161, 162, 260, 262
post-translational, 111, 156, 212
post-translational modifications, 156, 212
potato, 114, 117
power, 5, 25, 26, 27, 29, 34, 190, 205, 217, 243, 256, 266, 287
preconditioning, 132
prediction, 56, 261
predictive model, 30, 73, 87, 100, 101, 112, 158, 179, 181, 190, 206
preprocessing, 75
president, xii, 1, 4, 172, 289, 296, 299, 303
pressure, 15, 66, 86, 147, 166, 184, 203, 218, 225
prices, 10, 172, 266
primary products, 210
private, 25, 29, 98, 103, 108, 109, 164, 239, 301, 306
private sector, 25, 164, 239
probe, 73, 128, 148, 234
process control, 211
producers, 145, 214, 255, 260, 267, 269
production costs, 145, 249
production function, 104
productivity, xiii, 4, 14, 15, 19, 44, 52, 69, 81, 86, 95, 97, 98, 99, 100, 101, 102, 103, 104, 112, 124, 172, 176, 177, 178, 181, 183, 185, 205, 213, 223, 255, 303
progenitors, 175, 225
progeny, 93, 94
program, xv, 4, 6, 27, 29, 32, 74, 82, 93, 108, 109, 116, 131, 161, 199, 204, 218, 222, 242, 245, 247, 259, 265, 267, 269, 282, 286, 287, 290, 291, 296, 297, 305
programming, 300
promoter, 61, 181
property, vi, xiii, 55, 184
prosperity, 9
protection, 3, 65, 199, 292
protein engineering, 136, 139, 141, 148, 211, 212, 283, 292, 297, 306
protein folding, 297

protein function, 164, 226
protein structure, 211, 213, 218, 231
protein synthesis, 198
protein-protein interactions, 262
proteoglycans, 59
proteome, 90, 111, 196, 201, 215, 226, 252
proteomics, 45, 92, 140, 159, 173, 185, 187, 203, 206, 223, 228, 242, 252, 253, 267, 288, 294
protocols, 129, 286
prototyping, 295
pruning, 183
public, 103, 268
public health, 268
pumps, 25, 189, 214, 217, 246
purification, 70, 200, 214, 253, 292, 295
pyrolysis, 73, 74, 160, 217

Q

quality assurance, 291
quality control, 291
query, 239

R

R&D investments, xii, 1
radiation, 281, 285
rain, 95
rainfall, 87, 89, 95, 115
eaman, 128, 231, 282
Raman scattering, 231
Raman spectra, 282
raman spectroscopy, 282
random, 136, 153, 298
randomness, 137
range, xiii, 2, 33, 44, 46, 50, 51, 52, 56, 72, 81, 82, 127, 130, 131, 133, 134, 135, 142, 147, 149, 159, 162, 178, 189, 199, 204, 227, 230, 233, 235, 237, 240, 242, 247, 267, 289, 298
rapid prototyping, 295
raw material, xv, 259, 270
reaction mechanism, 284
reaction rate, 125
reactivity, 161
reading, 243
reagents, 73, 74, 92, 189, 249, 250
real time, 73, 112, 151, 233
recognition, 163, 243
reconstruction, 153, 187, 208, 296

recovery, 18, 115, 165, 188, 210, 213, 214
recovery processes, 213
redox, 181, 183, 190, 207, 213, 245, 253
refineries, 2, 8, 18
refining, 44
regional, 26, 52, 53, 300
regular, 7
regulation, 26, 32, 33, 34, 47, 50, 62, 72, 75, 94, 125, 134, 175, 179, 181, 188, 189, 193, 195, 205, 207, 208, 210, 211, 218, 222, 227, 245, 253, 301, 303
regulators, 211, 212
regulatory controls, 52, 155, 179
relationships, 35, 62, 66, 74, 129, 130, 136, 139, 142, 144, 148, 160, 161, 218, 244, 261, 284, 296, 301, 306
relevance, 211, 303
renewable energy, 5, 25, 142, 285
Renewable Fuel Standard, ii, 267
renewable resource, 288, 295
repair, 281
replication, 137
reporters, 238
reproduction, 99, 203
research and development, 6, 20, 21, 36, 37, 53, 175, 219, 255, 265, 266, 274, 288
reserves, 9
residential, 296
residues, xiv, 3, 10, 11, 16, 17, 18, 19, 26, 44, 56, 60, 67, 83, 91, 97, 100, 102, 123, 145, 156, 159, 224, 293, 304
resistance, 14, 21, 51, 64, 89, 93, 108, 194, 195, 199, 211, 213, 215, 222
resolution, 73, 126, 160, 163, 227, 231, 233, 234, 235, 236, 237
resources, 3, 4, 11, 14, 15, 18, 23, 25, 56, 66, 92, 104, 105, 110, 112, 113, 114, 131, 142, 155, 157, 198, 204, 209, 212, 215, 216, 224, 240, 249, 253, 255, 256, 268, 288, 295, 305
respiratory, 206, 289, 291
responsibilities, 282, 286, 305
reticulum, 111, 297
retirement, 303
revenue, 9, 10
RFA, 9, 20, 37
rhizosphere, 101, 112, 135, 151
rice, 57, 89, 98, 104, 110, 156, 274, 293, 300
rigidity, 285
risk, xv, 3, 6, 7, 44, 160, 172, 175, 188, 259, 304
RNA, 157, 225, 227, 230, 231, 252

roadmap, xiii, 2, 23, 32, 33, 38, 43, 160, 172, 174, 179, 219, 236, 240, 241, 252, 254, 257
robotic, 160, 163, 164, 229, 250, 251, 293
robustness, 21, 45, 51, 70, 159, 165, 175, 177, 188, 203
royal society, 170, 301
rubber, 268
ruminant, 202, 203, 303, 306
runoff, 85
rural, 10, 267

S

salinity, 52
salts, 161
sample, 74, 160, 162, 200, 229, 234, 235, 237
sampling, 227, 229, 254
savings, 7, 192
sawdust, 16
scaffold, 63
Scanning Electron Microscopy (SEM), 58, 126, 128, 234, 235
scattering, 231
scientific community, 28
scientific progress, 68
search, 143, 154, 215
searching, 148
Secretary of Agriculture, 6, 269
secrete, 210
secretion, 140, 144, 149, 197, 212, 237
security, 3, 6, 7, 8, 9, 24, 25, 28, 29, 51, 124, 267
seed, xiii, 81, 84, 114, 117
seedlings, 91
selecting, 159
selectivity, 233
self, 209
self-assembly, 154, 283
semicrystalline polymers, 282
senescence, 166
sensitivity, 75, 160, 162, 163, 195, 227, 229, 233, 246, 260, 287
separation, 22, 165, 192, 199, 210, 213, 227, 228
sequencing, 47, 87, 89, 93, 98, 101, 102, 105, 107, 108, 114, 151, 157, 185, 199, 204, 211, 215, 222, 224, 243, 247, 248, 297, 305
series, 283, 299
services, vi
severity, 22, 33, 132, 143
sexual reproduction, 99
shape, 129, 281

sharing, 239
short-term, 224
signal transduction, 286
signaling, 158, 203, 228, 305
signals, 204, 243, 289
silica, 98
similarity, 67, 255
simulation, 29, 31, 32, 47, 75, 90, 147, 173, 205, 223, 237, 241, 244, 245, 251, 284, 299
single nucleotide polymorphism, 92
sites, 103, 125, 128, 138, 189, 207, 224, 235, 253, 269
sludge, 296
SNPs, 92, 93, 98, 107
SOC, 100
social impacts, 3
social institutions, 26
social systems, 24, 26
software, 132, 234
soil, xiii, 11, 18, 19, 44, 49, 53, 81, 82, 83, 84, 85, 87, 89, 92, 95, 97, 99, 100, 101, 102, 103, 104, 111, 115, 135, 157, 203, 216, 223, 224, 238, 243, 247, 255, 290, 292, 306
soil erosion, 85, 92
solar, 25, 85, 114, 282, 289
solar cell, 282
solar energy, 114, 289
solid waste, 16
solid-state, 282
solubility, 66, 67, 211
solvents, 73, 189, 217, 288
soy, 116
soybean, 18, 95, 106, 114
spatial, 158, 162, 163, 204, 208, 209, 231, 235, 237, 246, 249, 251
specificity, 50, 52, 61, 63, 72, 110, 172, 307
spectroscopic methods, 160
spectroscopy, 113, 128, 129, 141, 190, 227, 235, 251, 287, 297
spectrum, 291
speed, xi, 56, 72, 82, 165, 181, 231, 235
stability, 44, 133, 139, 140, 233, 281
stabilize, 203, 285
stages, 27, 94, 127, 159, 161, 210, 229
standards, ii, 70, 73, 113, 142, 164, 167, 229, 239, 285, 291
starch, 4, 20, 21, 86, 94, 114, 116, 143, 145, 146, 157, 161, 171, 172, 214, 297
State of the Union, 1, 36
steady state, 200, 211
steel, 63
sterilization, 195, 198
stock, 105
stoichiometry, 147, 207
storage, xiv, 66, 69, 86, 94, 95, 114, 115, 116, 123, 129, 164, 167, 235, 242, 299
strains, 9, 20, 21, 47, 50, 52, 124, 134, 140, 148, 175, 176, 178, 179, 180, 181, 183, 186, 187, 188, 192, 195, 201, 203, 213, 223, 225, 243, 255, 261, 286, 294, 300, 305
strategic planning, 6, 281
strategies, xi, xii, xiv, 9, 14, 23, 25, 45, 62, 104, 109, 123, 140, 160, 163, 165, 173, 180, 194, 196, 205, 212, 224, 234, 254
streams, 161, 165, 176, 198
strength, 21, 63, 64, 66, 293
stress, 15, 34, 50, 53, 59, 90, 97, 108, 175, 190, 207, 212, 244, 256
stressors, 52
structural changes, 229
structural protein, 189
students, 304
substances, 51, 192
substitutes, 266
substitution, 136
subtilisin, 297
Succinic, 288
sucrose, 161
sugar, 20, 26, 33, 34, 45, 50, 52, 56, 67, 110, 114, 125, 127, 140, 142, 145, 165, 172, 177, 179, 180, 183, 189, 214, 215, 222, 225, 227, 230, 232, 250, 253, 260, 269, 284, 290, 293
sugar beet, 114
sugarcane, 95, 269, 293
sulfur, 11, 217
summer, 302
suppliers, 162
supply, 5, 6, 9, 13, 14, 15, 18, 25, 33, 82, 100, 199
supramolecular, 151
surplus, 86
survival, 86, 203, 216
surviving, 215
susceptibility, 126
sustainability, 3, 19, 33, 49, 69, 86, 90, 91, 99, 103, 104, 115, 225, 248, 267
symbiosis, 292
symbiotic, 292
synchrotron, 235
synergistic, 31, 125, 143, 144, 147, 148, 191, 211

synthesis, xiii, 44, 59, 60, 64, 68, 69, 70, 72, 74, 75, 81, 106, 110, 111, 112, 116, 124, 164, 181, 198, 206, 218, 228, 229, 244, 249, 250, 253, 255, 256, 291, 292, 294, 298, 301, 303

T

talent, 299
tangible, 86
tanks, 192
targets, 5, 22, 43, 144, 178, 185, 193, 201, 225, 286
task force, 300
teaching, 288, 298
technological progress, 296
technology, xiii, 3, 4, 26, 29, 31, 43, 44, 48, 51, 56, 57, 115, 125, 145, 163, 165, 166, 173, 174, 177, 191, 200, 209, 214, 217, 224, 231, 257, 262, 281, 282, 283, 285, 289, 290, 293, 299, 304
temperate zone, 85
temperature, 50, 52, 66, 89, 130, 147, 156, 166, 215, 217, 218
temporal, 95, 158, 159, 237, 251
tension, 285
termites, 225
tetrahydrofuran, 136
textile, 293
thermal stability, 140
thermodynamic, 144, 147, 148, 149
thermophiles, 188, 191, 197
thermoplastics, 298, 299
threat, 8, 98
three-dimensional, 149
threonine, 183
time, 11, 26, 72, 82, 93, 94, 100, 104, 105, 111, 112, 127, 130, 140, 143, 146, 162, 166, 193, 196, 203, 210, 223, 230, 235, 236, 252, 254, 261, 288, 291, 294, 296
time consuming, 93, 162
time periods, 111
tissue, 56, 57, 61, 63, 65, 67, 75, 91, 115, 256
title, 285
tolerance, 15, 33, 34, 47, 49, 50, 52, 53, 82, 86, 93, 97, 108, 127, 146, 175, 177, 183, 184, 185, 187, 188, 189, 192, 193, 195, 196, 207, 208, 215
toxic, 45, 49, 51, 67, 194, 208, 289
toxic substances, 51
toxicity, 195
toxins, 71, 124, 207, 281
tracers, 183
tracking, 229, 241, 249

trade, 9, 293
trade deficit, 9
traditional paradigm, 285
training, 284, 298, 300, 302
traits, xiii, xiv, 14, 33, 34, 45, 46, 52, 53, 81, 82, 89, 92, 93, 94, 97, 98, 99, 104, 105, 106, 108, 123, 124, 175, 180, 184, 187, 189, 191, 193, 195, 200, 201, 216, 222, 225, 262, 282
transcript, 108, 190, 203, 227, 253
transcriptional, 70, 108, 183, 185
transcriptomics, 92, 159, 203, 223, 225, 228, 252, 291
transducin, 207
transduction, 289
transfer, 34, 50, 58, 110, 136, 151, 292
transformation, xiii, 4, 56, 73, 92, 93, 99, 105, 107, 108, 110, 112, 113, 132, 166, 184, 205, 218, 245, 295, 305
transgenic, 34, 61, 70, 82, 109, 119, 120, 141, 305
transition, 3, 82
transmission, 128, 235
Transmission Electron Microscopy (TEM), 128, 234, 235
transpiration, 285
transport, 25, 63, 65, 75, 95, 115, 130, 206, 207, 227, 244, 246, 289, 292
transport processes, 207
transportation, xiii, 2, 3, 5, 6, 7, 9, 11, 13, 14, 15, 20, 21, 27, 35, 36, 37, 40, 41, 43, 44, 99, 165, 213, 266, 269, 270
transportation infrastructure, 213
trees, xiv, 18, 61, 72, 90, 94, 104, 112, 123, 141
trimmings, 16
tubers, 114
tunneling, 233
turnover, 230

U

U.S. Department of Agriculture (USDA), 5, 10, 14, 15, 36, 37, 56, 82, 98, 101, 108, 109, 113, 117, 156, 273, 274, 275, 276, 277, 278, 280, 282, 283, 287, 293, 298, 305
U.S. economy, 3
ubiquitin, 299
ultrastructure, 57, 123, 131
undergraduate, 295, 298, 302, 306
underlying mechanisms, 87, 212
universe, 148
universities, 29, 98, 290

V

vacuum, 234
validation, 160, 163, 164, 173, 246, 248
values, 86, 125, 261
variability, 109, 304
variables, 74, 204, 254
variation, 73, 93, 94, 95, 200
vehicles, 3, 15, 266, 281, 296
versatility, 159, 234
vision, 6, 37, 72, 112, 113, 118, 159, 246
visualization, 70, 189, 204
vitamins, 290, 305

W

waste management, 283
waste treatment, 203
wastes, 285, 306
wastewater, 165
wastewater treatment, 165
water, 66, 95, 97, 126, 130, 131, 143, 150, 176, 188, 213, 214, 216, 244, 268
water resources, 268
wavelengths, 231

waxes, 114, 161
wealth, 243
web, xv, 259
wheat, xiv, 18, 28, 95, 123
white house, 1, 35
wilderness, 18
wildlife, 11
wind, 25
winning, 292
wood, 5, 16, 17, 21, 44, 51, 74, 104, 106, 159, 293, 300, 304
wood waste, 293

X

xylem, 65

Y

yeast, 20, 21, 46, 151, 177, 180, 183, 185, 186, 187, 199, 215, 223, 255, 300

Z

Zea mays, 85, 120